'There is a huge stack of books about ⸻ tedious and repetitive. This is the op⸻ ⸻ engaging account of the most important task of our time, convincing the world to act on what science tells us before it's too late. It's funny, warm, human, sympathetic—and tough-minded too!'

Bill McKibben, author of *Eaarth* and *The End of Nature*

'This is a first class read. I love Anna Rose's passion. Ironically, experiencing the moral and intellectual stubbornness and intransigence of the likes of Minchin should serve to open the hearts and minds of many others.'

John Hewson,
federal leader of the Liberal Party of Australia, 1990–1994

'Anna Rose is perceptive and passionate. This inspiring book is about the divide between Australians. The developing world needs action on climate change. If we fail, it will condemn their development and thwart their attempts to feed their citizens.'

Tim Costello, CEO, World Vision Australia

'I can't recommend this book highly enough . . . I loved it and couldn't put it down. An entertaining, sincere and highly educational piece of work that I hope will find its way into every Australian's hip pocket . . . ASAP.'

Natalie Pa'apa'a, lead singer, Blue King Brown

Anna Rose is co-founder and chair of the Australian Youth Climate Coalition, a movement of over 70,000 young people working to solve climate change before it's too late. She is the joint recipient of the 2009 Environment Minister's Young Environmentalist of the Year Award and a Fellow of the International Youth Foundation.

Anna's climate change work started with shovelling compost and emptying recycling bins in high school. She went on to work with the National Union of Students on dozens of campus clean energy victories, speak onstage with the Dalai Lama to an audience of 16,000, trudge through snow door-knocking in the United States for the Obama campaign, and spearhead Australia's first climate torch relay in her former role as GetUp's climate campaigner.

In 2011 Anna received a Churchill Fellowship to research youth peer-to-peer climate education. The *Sydney Morning Herald* named her one of Sydney's most influential people (in 2009) and one of the top five most powerful grassroots organisers in New South Wales (in 2011). She lives in Sydney with her husband Simon and the occasional foster cat.

MADLANDS

A JOURNEY TO CHANGE THE MIND OF A CLIMATE SCEPTIC

ANNA ROSE

MELBOURNE
UNIVERSITY
PRESS

10% of author royalties are donated to the Australian Youth Climate Coalition
www.aycc.org.au

MELBOURNE UNIVERSITY PRESS

An imprint of Melbourne University Publishing Limited
187 Grattan Street, Carlton, Victoria 3053, Australia
mup-info@unimelb.edu.au
www.mup.com.au

First published 2012

Cover design by Design by Committee
Text design by Bookhouse, Sydney
Typeset by Bookhouse, Sydney
Front cover photos of Anna and Nick by Kate Hodges and Max Bourke, Smith & Nasht
Author photo by Milena Dekic, I Love Wednesdays
Printed by Griffin Press, South Australia

National Library of Australia Cataloguing-in-Publication entry:

Author: Rose, Anna.
Title: Madlands : a journey to change the mind of a sceptic / Anna Rose.
ISBN: 9780522861693 (pbk.)

Subjects: Climatic changes--Environmental aspects.
          Climatic changes--Effect of human beings on.
          Climatology--Observations.
          Global warming.
          Skepticism.

Dewey Number: 551.6

Written in memory of my friend David Arkless.
And in awe of the 70,000 young change-makers
in the Australian Youth Climate Coalition—
you carry on the legacy David and I began in high school
when we set up the Merewether Greenies together

CONTENTS

CONTENTS

Chapter One

MINCHIN IMPOSSIBLE

I'M WAITING ON THE TARMAC AT MOREE AIRPORT IN NORTH-WESTERN New South Wales as Leo hurriedly attaches a microphone to a velcro belt around my waist. He's one of the best sound guys in the business and it's crucial that he gets audio for the strange scene that's about to take place. The director, Max, pushes me towards the small plane to greet the man with whom I'll be spending the next four and a half weeks. I'm about to meet the Liberal Party powerbroker and former senator Nick Minchin.

Nick's plane from Sydney is early. I had to stop and wait for some cattle on the road from my uncle's farm this morning so I'm running late. I approach the plane that's just landed. The propellers are still spinning. Nick walks down the steps and onto the tarmac and looks around expectantly. 'Welcome to Moree,' I say and extend my hand to shake his. I'm genuinely happy to finally meet him after several days of setup shots with the film crew and many weeks of planning how I'll approach this unique project.

Nick Minchin cuts an impressive figure. Medium build, with thin-rimmed glasses and a full head of steel grey hair, he has the manner of an experienced politician. He's confident, articulate and comfortable under the gaze of the camera. He has a focused and intense stare, but he breaks easily into a smile as he walks towards me.

It had been a long journey leading up to this meeting. I'd first met with the documentary producer, Simon Nasht, a few months beforehand. Simon wanted to talk to me about an idea he and his business partner, Aussie entrepreneur Dick Smith, had for a documentary project to be screened on ABC TV. He wanted to create a program that reached beyond the sound bites on climate change science. His vision was to send a film crew to capture the journey of a climate activist and a climate sceptic as they took each other around the world trying to change each other's minds.

Simon Nasht had secured one of the remaining few high-profile climate sceptics in Australia, Nick Minchin, to be part of the project. (I say remaining because many other former sceptics in business and government, even the former head of Exxon Mobil, John Schubert, have now accepted the science.) Now, Simon required someone willing to go head to head with Nick to argue the case for the science and the need to act on climate change.

I was crystal clear about the fact that climate change is real and human-induced. Frankly, I was baffled by people like Nick. It seemed like climate scepticism (or climate denialism, as it's often called) rested on a refusal to accept basic principles of physics and chemistry as well as disputing the observational evidence. I'd learned about climate change in primary school—how the greenhouse effect worked and what the consequences were of burning fossil fuels that added heat-trapping gases to the atmosphere. That had been sixteen years ago. At the same time I'd witnessed the impacts of climate change first-hand after seeing unprecedented drought affect my grandparents' and uncle's farms in north-west New South Wales.

The realisation that humans were altering the earth's climate and that it was hurting the people and land I cared about spurred me into action. Now twenty-eight years old, I'd spent over a decade as a climate activist working with young people and doing everything I could to reduce Australia's carbon pollution.

I was interested to see what points the sceptics would raise if I did the documentary. I knew people like Nick were absolutely certain they were right—what was motivating them? More importantly, I wanted to put myself in their shoes to see if there was anything I could say or do that might prompt them to reconsider. Still, I was initially wary. I certainly didn't say yes to the project straight away. I wanted to make sure that my mind really was open to being changed—after all, science is all about continually testing your assumptions through inquiry. I thought it over with a mix of anxiety and excitement for several weeks and many sleepless nights.

Nick Minchin wasn't someone to take on lightly. He was a legendary powerbroker with a controlling stake in the right wing of the Liberal Party. He'd just been the subject of an article by the investigative journalism site *The Power Index* which had named him the seventh most powerful 'political fixer' in Australia. Dubbing him 'the Liberal Party's spiritual leader', the article lauded him as 'the Leader of the Right and the keeper of the Howard flame'. And even though Nick had just retired as a senator, he was still a formidable opponent and most commentators agreed he was still pulling strings from behind the scenes. Nowhere were those strings being pulled more enthusiastically than in the campaign to oppose cuts to carbon pollution and the flat-out rejection of the findings of climate science.

Nick had spent much of his life staunchly refusing to accept any link between the carbon pollution emitted by humans and the climate change the world was already experiencing. He'd also influenced a generation of young Liberal Party warriors who were busy moulding themselves in his image. If it were possible to change his mind, it would be an enormous victory for climate science. I knew it was unlikely. But part of me held on to a tiny sliver of hope that exposing Nick to the overwhelming evidence and finding the right experts to explain it might just be enough to trigger a shift. Failing that I could reach the television audience—a large number of people open to learning more about the science. Perhaps participating in this project could help give some crucial information to people who genuinely weren't sure.

But I also knew there were risks.

I'd consulted colleagues in the range of environmental organisations and community groups that together make up the Australian climate movement. They were split on the question of whether or not I should participate in the project. Many were supportive and welcomed the opportunity to expose the weakness of climate sceptic arguments. However, I'd received strong advice from others to have nothing to do with the project. There were indisputable dangers: if I stuffed it up, a lot of good people's reputations would be on the line. The credibility of the organisation I'd co-founded, the Australian Youth Climate Coalition, was at stake. My performance would end up reflecting on the people around the country volunteering their time as part of the movement to avoid dangerous climate change. The reputations of scientists and experts I would secure as interviewees—in large part because they trusted my judgment—would also be on the line. It was a high-pressure decision.

People told me that they thought I was naive to think I could 'win' the argument; that the whole idea of the show played into the denialists' strategy of framing the science as disputed when it actually wasn't and that it was just an opportunity for Nick to place figures opposing the science into the mainstream media. Some told me the denialists would be celebrating a great coup in the fact that the show was being planned. An email from Cindy Baxter, a climate communications strategist and co-author of the website ExxonSecrets.org, which is run by Greenpeace, summed up the major concerns:

> It's a shame this is going ahead. The whole premise of the climate denier industry's strategy is that, like the tobacco industry's tactics before them, 'doubt is their product'. Their main tool has been to raise the possibility that the climate science is in doubt. They do this by getting programmes like this one to go ahead. You'll never be able to convince Minchin of your point of view. The whole premise of this program is a win for the deniers because it continues the meme that the main conclusions from the climate science are not true. Sorry to be so blunt, but having watched these guys for 20 years, I am only too aware of the game they are playing and the tactics they use to win. And how they win is by getting prime time TV to do this sort of programme. This is not a scientific argument, it's a political one where, for 20 years, the denier campaign has been funded by the fossil fuel industry whose product is causing the problem.

I agreed that there were reasons to be concerned. Virtually all scientists working and publishing in the field of climate change—whether atmospheric physicists, biologists, oceanographers, glaciologists or geologists—agree that climate change is happening and caused by humans. The format of the documentary, however, would frame the debate as a 50:50 one. I'd heard a rumour that the scientist

Tim Flannery and the ABC's science reporter Robyn Williams had already said no to participating.

Despite this, I decided to say yes. What convinced me was the fact that the program was going ahead anyway. It was clear that the production team would find someone willing to debate Nick if I said no. I also knew that many Australians were confused about the science of climate change and could do with some more information. I was confident I'd be able to change some hearts and minds in the audience. This was not a project to mobilise the base of people who already cared about climate change. This was a chance to reach a large number of people—directly in their living rooms—who the climate movement had struggled to communicate with for years.

The terms were set. I could take Nick to meet scientists and experts in Australia and overseas who I thought could change his mind on climate change. He could do the same, and try to change my mind by introducing me to his picks of sceptics. Simon Nasht, the executive producer of the program, would also choose some 'neutral' spokespeople who he felt had something useful to say to both of us. We'd film for four and a half weeks consecutively, travelling together with a film crew: Max the director, Kate the producer, Leo the sound guy, and Pete the cameraman. It was reality television, ABC style.

For a long time, many environmental groups and scientists tried ignoring the climate sceptics. The argument had traditionally been: 'If we tackle them head on, they'll get more airtime.' Others had been willing to challenge sceptics by exposing their sources of funding and the dirty tactics they used, such as the sending of threatening emails to climate scientists and activists. But that strategy only worked for so long and, by 2011, climate sceptics and deniers were dominating talkback radio, the daily tabloids, and our national newspaper, *The*

Australian. Why? It was an open secret. Mining and other carbon-intensive companies had joined with right wing think tanks and allies in the main political parties to orchestrate a well-oiled campaign to undermine climate science. The attack on climate science had been underway globally since the mid 1990s. It was relentless and effective. In 2006 one of the oldest and most venerable scientific organisations in the world, the UK's Royal Society, was so appalled by the situation that it wrote to Exxon Mobil asking them to desist from funding anti-science groups.

The year 2006 helped shed light on the people and organisations opposing responsible climate action. In Australia, former Liberal Party insider Guy Pearse appeared on the ABC's *Four Corners* program as a whistleblower. He went on record exposing the activities of Australia's top fossil fuel lobbyists. Calling themselves 'the Greenhouse Mafia', these lobbyists had essentially written Australia's greenhouse policy under the Coalition government since 1998, claimed Pearse. Despite these new insights into the anti-science movement, climate denialism hadn't slowed down. Instead, it shifted from an amply funded lobbying campaign to a wider movement. 'It can no longer be understood as lobbying funded by the fossil fuel companies but has become linked to a populist political movement,' author Clive Hamilton said at the launch of his book, *Requiem for a Species*. Features of this movement in Australia include a hive of aggressive blogs and websites, campaigns of cyber-bullying directed at climate scientists, and partnerships with groups holding a broader agenda of gripes against 'big government'. A journalist friend of mine calls this nexus 'an enormous, gargantuan dirty machine'. I just call them the 'anti-science forces' since their mission is to cast doubt on scientists and their work. They are unfortunately quite good at doing at this. Professor David Karoly from the University of Melbourne points out that between 1000 and 2000 peer-reviewed scientific papers are published each year on climate change. Not one of them contradicts the assertion that climate change is happening and caused by human activities. Yet the level of confusion and misunderstanding of climate science among the Australian public is, according to most opinion polls, at a record high.

Anti-science forces had been growing in influence since the 2009 Copenhagen Climate Conference. I went to Copenhagen to lead a delegation of twenty young Australians and fifteen young Pacific Islander climate activists. (If you ever want to feel your heart break in two, listen to a young Pacific Islander talk about what climate change is doing to their homelands and culture.) I'd felt the blow deeply when the talks collapsed. As the delegates passed nothing more than a weak two-page agreement, I was outside the gates of the negotiation halls protesting side by side with hundreds of other young people. 'It's our future you decide,' we'd chanted, many of us in tears, in the freezing night air.

This outcome wasn't inevitable. A combination of factors had derailed the political will of the negotiators. This included the media beat-up about thousands of private emails stolen by the anti-science forces from the University of East Anglia's Climate Research Unit. The hackers had alleged that some climatologists manipulated data, dubbing the whole situation 'climate gate'. Scientists were later cleared of having done anything wrong, but headlines had already reverberated around the world. Attack after attack like this has led to the previous broad consensus on climate change science among Australians being shaken.

•

My preparations for the project had consumed every waking hour outside work for weeks. My research books and notes took up several shelves. I didn't study science formally past Year 10 high school. I even staged a walkout of one of my science classes when the teacher wanted us to dissect mice. The science teacher said, 'If you're lucky, you'll get a pregnant one.' I was morally outraged, and aware that there were alternatives to killing mice for high school student experiments. So I stridently led a group of other conscientious objectors into the corridor to sit and, well, object, instead of participating in cutting up the mice. I stopped studying science at school after that. So when I try to communicate climate science I'm obviously relying on the work of other people: the CSIRO, the Australian National Academy of Sciences, NASA, the Intergovernmental Panel on Climate Change, and dozens of expert scientists. It's their thorough reports that help me appreciate the science.

In order to participate in the documentary I had to brush up on the most recent scientific papers. Then I had to research and select spokespeople who I felt had a fighting chance of changing Nick's mind—or at least of explaining things in a way the audience would understand. Finally, I needed to negotiate five weeks leave from my job at a communications agency for environmental and social justice groups. More difficult was negotiating time away from my fiancé, Simon, who runs the national progressive online advocacy organisation *GetUp*. I knew I'd miss him.

Simon and I were supposed to be planning our wedding in the time I'd be away filming. It had been almost a year since his romantic proposal down on one knee in The Rocks in Sydney. Initially we thought we'd have plenty of time. After we locked in the wedding venue we'd been pretty relaxed about the rest of the details. It would all somehow come together in the months before the big day, we'd figured. The problem was I'd only just returned to Australia from a two-month work trip to the United States, England and China. If I agreed to do the documentary it would mean I was going away again—this time returning only two and half weeks before the wedding.

We were both slightly alarmed at the prospect of organising a do-it-yourself wedding for 130 people in a national park in only a couple of weeks. However, Simon was—as always—incredibly supportive. He said that if I wanted to go, we could plan the wedding via Skype and phone. I decided that would be doable. Easy, even! So I created a Google spreadsheet called 'open source wedding'. I added my name and Simon's name next to some major tasks, and left the rest of the cells blank for our friends and family to fill their names in next to jobs that they wanted to help with.

With all that organised, I now find myself on the tarmac in Moree meeting Nick Minchin. I'd be lying if I pretended I wasn't nursing a healthy dose of trepidation and even fear. We shake hands and smile at each other. I know right away that this journey will be one of the hardest things I've ever done.

Chapter Two

TROUBLE ON THE FARM

THIS ADVENTURE IS ABOUT TO TAKE US ALL AROUND THE WORLD. BUT our first stop is Moree, New South Wales, in the big sky country west of Tamworth and north of Narrabri. I'm taking Nick to my Uncle Geoff's farm to show him why I learned to care about climate change in the first place. I didn't develop an interest in global warming because of polar bears or melting icecaps. For me, it was always about the land, the food it sustains and the human beings it feeds.

I grew up in the 1980s. I believe that many people intuitively understand that the climate has changed since then. They've watched the temperature record soar and they've seen increases in extreme weather events on television. They understand that pumping 30 billion tonnes a year of carbon pollution into the atmosphere has consequences. But our natural instincts can easily be swayed when we're hearing from the media that there's still a 'debate'. Who wouldn't rather relax and hope that our climate system will be OK? It's certainly more fun than listening to the warnings telling us we're at the start of major human-induced climate disruption.

I used to be oblivious to these warnings too, even though I was born in the belly of the beast of the climate change problem: the world's biggest coal export port. Newcastle is an industrial city with a deep water harbour at the mouth of the coal-filled Hunter Valley. My mum was born and raised on a dairy farm outside one of the valley's towns, Singleton. The Singleton Shire Council area isn't large, but it produces 57 million tonnes of coal from thirteen mines each year. Some is burned for electricity in the six power stations scattered throughout the valley. The rest is exported to Japan, India and China in an endless stream of coal ships. I used to see those ships when I'd surf in the mornings before school with my best mate, Angeline. We'd sit out on our boards watching the sunrise, silently observing twenty to thirty enormous ships sitting low on the horizon waiting to come into our harbour.

I didn't spend all my time in Newcastle though. Most school holidays I could be found on my grandparents' farm in Gunnedah. My grandfather was a farmer his whole life, as was his father. He'd drilled into me that farmers have three tools of trade: water, soil and climate. He also taught me that you don't 'own' the land. You belong to it. I was never allowed to forget that farming ran in my blood. During the Christmas break, my parents would drop me and my sister off at the farm and we'd run wild for weeks on end. We'd chase each other through paddocks of wheat, tumble on and off horses and climb the majestic river red gums beside the creek. Long summers always included camping, jumping off hay bales in the shed and learning to help with the cattle and the crops. There was tragedy—like the time a brown snake killed our dog. And there was triumph—every new piece of information we gleaned about the natural world.

I learned that 'the environment' isn't something pretty that you visit on weekends in order to unwind, as the television ads for nature getaways imply. The environment is what we depend upon for our food and water, and therefore our survival.

As I grew older, the drought worsened. At one stage it had been so long since it had rained there were toddlers in parts of New South Wales who didn't know what rain was. My grandparents' farm, Manaree, and my uncle's farm, Teralba, were three hours drive away from each other, but were both getting harder to farm.

At the same time, I was learning about the greenhouse effect in primary school. We were drawing diagrams of the sun, the atmosphere and the carbon cycle. My primary school was Catholic, with a strong social justice bent. We had a 'sponsor a school' program for which the kids brought in gold coins every Friday. But our donations weren't going to a school in Africa. Instead, we were supporting a school in a drought-stricken town in northwest New South Wales. Our gold coins were helping pay for water to be trucked in to the town of Manilla, where the water was running out.

My grandparents sold their farm in Gunnedah. Drought was a factor, although not the only one. They were getting older and farming wasn't getting any easier. My uncle kept his farm in Moree but sent a lot of his cattle away temporarily to farms that were less crippled by drought. Farmers waited for rain. The land changed as the drought continued on and off. It seemed to worsen as I graduated from high school in 2001. In the summer of 2002–2003 eighty-eight per cent of the state was declared under 'exceptional circumstances' due to drought. But it was no longer exceptional to those of us who'd spent time on the land. It was the new normal.

The drought has now broken in many places in Australia, replaced in some areas by extreme floods. In light of all this, I often worry about the future of Australian farming. I worry that forty per cent of

Australia's food is grown along the Murray–Darling basin, which is now almost completely degraded. I worry that the Garnaut Review (the study commissioned by the federal government into climate change impacts on Australia) told us that the Murray–Darling's agricultural production could decline by ninety-two per cent by 2100 if no action is taken to reduce carbon emissions. This decline could be as much as ninety-seven per cent in a hot, dry, extreme case of climate change.

•

After driving out to the farm, Max wants to film me introducing Nick to my Uncle Geoff. Nick and I sit in the car outside the farmhouse while Max sets up the shot. Leo, the sound guy, attaches a lapel microphone to my uncle's short-sleeved blue cotton shirt. Leo is the oldest of the camera crew and goes about his work with a quiet professionalism. He's a man of few words but his kindness shines through. Max, the director, gets the cameras ready. He's about ten years older than me, tall, wiry and full of energy. Like a first-time father with a video camera, he wants to capture *everything* on film. The producer, Kate, chats to my uncle to make sure he understands what they want from this scene. She's the most organised person I've ever met, and in charge of all the logistical details for the coming trip. She's smart, fun and lovely—even offering to help me with wedding planning while we're on the road. Our cameraman, Pete, is overseas on another job. He'll meet us at Sydney airport after the first leg of filming is done. In the meantime a freelance videographer, Helen, is filling in.

Nick's phone rings while we're waiting outside the house. He apologises, answers it, and dives into a long conversation. It's clearly about some political matter involving the Liberal Party. It doesn't sound like he's retired. On the contrary, it's clear Nick enjoys politics, and is good at it.

After he finishes his phone call I take him to meet Uncle Geoff. They shake hands. Nick is friendly, charming and polite. He knows a bit about this part of the world. An old friend who used to board with him at high school lives in the area and Nick used to visit. He and Geoff chat affably about the crops, the weather and the array of solar panels powering the house from Geoff's roof.

We're here to talk about the impacts of climate change on Australian farming. Australia's leading expert on agriculture and climate change, Dr Mark Howden, is supposed to be with us in Moree. He's a highly respected—and busy—CSIRO chief research scientist. But at the last minute the production team changed their minds about involving him. Apparently I have too many spokespeople and it would be unfair to Nick. I'm disappointed. Mark is the kind of measured, careful scientist who knows almost everything there is to know about his subject area. He would also come across well on television. With his thick silver hair, he looks like he could be a distant relative of George Clooney.

I'd been to visit Mark in Canberra with Amanda McKenzie a few weeks earlier. Amanda co-founded and co-directed the Australian Youth Climate Coalition with me from 2006 to 2011. She is also one of my best friends and a bridesmaid at my upcoming wedding. She and I had spent a Saturday afternoon with Mark to learn more about climate change and agriculture, the three of us sitting in his small office at the back of his home in a quiet suburb of Canberra. As we talked, his children and their friends played in the garden with their dog. It set a strangely idyllic scene for the disturbing agricultural scenarios we were discussing.

At our meeting, Mark noted that Australian farmers are very adaptable—some of the most skilled in the world. However, he also pointed out that there are limits to adaptation if climate change continues unabated. Speaking at a recent conference about the potential for a 4 degrees Celsius increase in global average temperatures, Mark had warned the audience that climate change could cause Australia to become a net importer of wheat. Coming from a family of wheat farmers, this seems unthinkable to me. But already the food price index is higher than ever before. The value of Australia's fruit and vegetable imports is currently more than the value of our exports. Maybe our country's food security isn't as secure as city folk would like to assume.

·

Nick and I wait in the corner of the front paddock on the horses we've been riding all day. Uncle Geoff parks the old white ute under a tree while Max sets up the camera. Max doesn't want us to start talking until the sun is lower in the sky; it'll make a better shot. While we wait, my mind flips through the statistics and information I learned from Dr Howden. There's so much to remember. But my train of thought is soon interrupted as I glance sideways at Nick's horse. He's jolting his head up and down, leaning to one side and pawing at the ground—classic signs the horse is about to roll. Nick notices too. He quickly takes his feet out of the stirrups and jumps sideways just as the horse rolls joyfully in the dirt, saddle still on. I'm mortified. Nick could have been seriously injured if he hadn't moved out of the way in time. But after his horse stands up again, Nick jumps back on. He looks surprised but not shaken. I'm impressed. He's obviously made of tough stuff.

After this rather dramatic event, Nick and I ride to the middle of the paddock and dismount. We tie the horses behind us and they happily munch on the hay in the tray of the ute. We're in the ploughed front paddock, standing on black alluvial soil that's some of the richest in the Southern Hemisphere. There's a gum tree towering above us, and the sun has begun its slow descent. Against this serene backdrop, Uncle Geoff talks to Nick about the farm. 'I've been here about twenty-six years,' he says. 'I've always loved the land. I've always loved farming'.

'Why this country, why this property?' asks Nick.

'It's very rich country,' says Geoff. 'And it sort of just gets into your blood, I think, after a time.'

At the start of the conversation, Nick asks my uncle a lot of questions. But they're not about climate change. He asks what Geoff does in terms of conditioning the soil. He asks about the spacing between crops when they're planted. He asks about tractors. He asks about floods: how often they happen, which rivers they start from and whether Geoff stores the flood water. He asks about cane toads ('those bloody frogs'), about bore water and the level of the basin. He's certainly interested in farming but I'm really struggling to steer the conversation towards climate change. I know that Geoff has a lot to say on the topic, but I'm worried he won't get a chance before the sun sets.

Eventually, Max interrupts and asks if Geoff can talk about the differences he's seen on the land.

'Over the years I've noticed a few changes that have happened,' says Geoff. He starts by talking about food prices worldwide. 'In terms of the prices we're getting for our products . . . they're varying a great deal more than they used to,' he says. 'We're noticing really big spikes in prices because of dramatic climate events.' Geoff holds degrees in both agriculture and economics from the University of New England. Like all farmers, he closely monitors global food and fibre prices.

'I guess there are a lot of reasons for that?' asks Nick.

'There *are* a lot of reasons,' replies Geoff. 'But I don't think there is any doubt about the fact that dramatic climate change events, like floods and droughts, have an effect.'

Geoff talks about extreme weather events worldwide—droughts in Russia, floods in Pakistan and China, fires in the United States—and their impact on food prices. The increase in these kinds of events is consistent with the predictions of climate change models. The fallout is devastating and the impact on the production, price and availability of food is long-lasting. The wildfires that swept Russia in 2010 destroyed 25 million acres of land and forest. Twenty-six per cent of the nation's wheat crops were wiped out. The implications for food security were so severe that the Kremlin placed a ban on wheat exports. In October 2011, Thailand suffered the worst flooding in its history. The damage was equivalent to 18 per cent of its gross domestic product.

Author and climate change activist Bill McKibben points out in his 2010 book *Eaarth* that the most important question that's been asked throughout human history is 'What's for dinner?' As climate change worsens that question is getting harder and harder to answer for millions of people around the world.

Our conversation continues, jumping around from topic to topic as the sun inches closer to the horizon. Perhaps I'm being paranoid but it seems that Nick interrupts a lot, right when Geoff starts to make a really good point. I know from experience that this is standard practice for clever politicians. When your opponent is in the middle of saying something really clearly, interrupt so the sound bite won't be usable on television. Then again, maybe Nick is just really engaged in the conversation.

Geoff turns to another example of climate change manifesting on the farm. 'In the last couple of years, we've seen buffalo fly on the place,' says Geoff. 'They've never been seen here before. Never been seen in this district.'

Buffalo fly are blood-sucking creatures that cause serious production losses for farmers. They swarm in enormous numbers and prey on cattle. Imagine you're a cow in this position. You're not likely to just stay in the one place and eat; you're going to walk around a lot to try to get away from the flies, and rub up against other cows to get the flies off you. You do this instead of constantly chomping mouthfuls of food, so you lose weight. And if you're a skinny cow, your farmer isn't happy, because you're sold by the kilo. Suddenly the humble buffalo fly has made a real impact on farmers' bottom lines.

'They go for the cattle, do they?' asks Nick.

'Pretty savage on cattle,' says Geoff, shaking his head.

What does this have to do with climate change? Well, buffalo flies were never seen in the Moree district until the past few years. Geoff says in the past they were regarded as pests only in the northern regions of Australia—the tropics. But now, local farmers have to inoculate their cattle against them. Geoff says, 'It's right across the area and we're expecting it to be an annual problem.' Inoculation is expensive and needs to be repeated every year. 'It's going to be costly,' says Geoff.

The spread of buffalo fly is consistent with climate change predictions that the CSIRO made back in the late 1980s and early 1990s. As the climate changes, plants and animals—including pests—relocate. As climate zones shift, so do a whole range of species.

Nick doesn't dispute that climate change is causing issues for farmers like Geoff. But he argues that humans aren't the primary cause of these changes. I know that over the course of our journey he's aiming to convince me that it's just natural variability that's the cause, or the sun, or cosmic rays. I'm open to hearing the evidence for these explanations, as unlikely as they sound.

But for now, Nick makes a glib joke. 'It's plant food, mate,' he grins to Geoff. 'Without CO_2 there ain't no plants. They love it.'

This is something I realise is a common theme for those who reject the science and its implications. The 'plant food' line often comes up on the sceptical blogosphere. But Geoff points out that too much CO_2 leads to natural systems getting out of balance. 'Too much carbon dioxide and you've got no plants,' he says. 'We know that carbon dioxide is going to affect plants, therefore water, therefore our land—and all these things are interrelated . . . It's not a joking matter to me.'

If you're talking about a monoculture (a crop of just one type of plant) in a closed system (like a greenhouse) then pumping in more CO_2 can indeed make plants grow faster. But a real ecosystem presents a different story. With increased CO_2 you'll see the immense growth of only some plants, often weeds. These then tower above other plants and block them from absorbing sunlight, or take up valuable soil nutrients or space for growth.

Geoff moves on to his final points about how climate change has impacted the farm. He tells us that farmers decide the right time to plant their crops based on the temperature of the soil. Every morning Geoff measures the soil temperature using a thermometer at seed depth. When the soil reaches the crop's ideal temperature and is rising, he plants the crop. Geoff tells us that, over time, the soil is getting hotter earlier. This means he has to sow summer crops weeks earlier. 'You sense something is changing, do you?' Nick asks. 'I more than sense it,' says Geoff. 'Things like this indicate it.'

The kinds of crops Uncle Geoff can grow have also changed. Previously Geoff used to grow temperate grasses for cattle to graze on. Now that climate zones are starting to shift, he's sowing tropical grasses. These grasses originated from Kenya and used to grow only in northern Australia. You wouldn't call Moree a tropical place by any stretch of the imagination—but now that air and soil temperatures have risen, tropical grasses can flourish here.

Nick tries to find a bright side when Geoff makes this point. 'Maybe it's a good thing you can grow them here?' he asks. Geoff replies: 'It's more that we grow what we *can* grow,' replies Geoff. 'But things are changing here.' While in Geoff's part of the world farmers can simply replace one crop with another (for now) this isn't the case in other parts of the country and the world. As I write, parts of southwest Western Australia are experiencing the lowest rainfall since records began in 1900. Farmers there aren't 'adapting'—they're despairing.

Droughts come up, as they always do when talking about climate change in Australia. Crops don't do well in extreme heat and when deprived of water. Geoff says that changes in rainfall patterns are what really worry him. 'We've got fewer rainfall events,' he says. 'When it does rain, it rains a lot—but there are long dry periods in between.' The CSIRO's research shows that reductions in autumn rainfall are affecting farmers around Australia, especially in southern New South Wales, Victoria and southwest Western Australia.

Nick accepts that Geoff's observations are accurate. 'You're sensing some warming and . . . nobody disagrees,' he says. 'I accept, particularly for Australian farmers who live in a highly unreliable climate, that climate change is a very big issue. I accept that.'

I've seen previous statements where Nick has claimed the world is cooling—or at least not warming. But Nick appears to be starting this journey from the premise that temperatures *are* rising. 'The planet is going through a warming phase,' he says. This admission is a good first step but now he's stuck on the question of why—is it human emissions of carbon pollution, or something else? 'The core of this debate is not that the climate is changing but what is driving it and is there actually anything we can do about it,' Nick says to me and Geoff. 'We really need to know what the causes are if we are going to do anything about it.'

Geoff doesn't accept this line of argument about us not knowing the cause. 'Some things are very easily measured, like the amount of carbon dioxide in the atmosphere. We know it's going up dramatically,' he says. 'I just think it's foolishness, foolish of us not to listen to the best of science.'

As a farmer, Geoff has a pretty good idea about the importance of the work of scientists. 'Farmers are always looking for any research, or any information that's coming from research . . . because we depend on that,' he says to Nick. Farmers get important information from the CSIRO, the Bureau of Meteorology and other scientific bodies. It wouldn't make sense for them to ignore what these bodies say on any issue, whether it be climate change or crop disease.

'I'm a farmer,' says Geoff. 'And so I don't know all the research that is being done. But it's a reality and it concerns us because I've got a responsibility to my kids to be able to produce food. For me it's a real world thing . . . I'm a farmer, but first and foremost, I'm a father.'

•

I suspect that climate change may not really be about science to Nick. In an interview with ABC TV's *Four Corners* in 2009 he'd said that climate change was an opportunity for progressives to 'de-industrialise the Western world' and environmentalism was 'the new religion' of the extreme left. But it's neither of those things to me, nor to any of the climate activists I know and work with. For me, it's basic science that I learned about in primary school and has now evolved into a serious threat. Considering that almost all the world's scientists agree on the science, Nick's position makes me angry. It's not just his future he's risking with his views about the climate system. It's mine and everyone else's.

I often wonder what would have happened if John Howard had won the 2007 election. He had promised to introduce an emissions trading scheme. What if it had been the Liberal Party to bring in legislation to tackle climate change? Nick Minchin's views might have been sidelined. As I write this, the leader of the Liberal Party, Tony Abbott, has sworn a 'blood oath' to repeal the price on carbon pollution if he should be elected leader. The bipartisan support we once had on responsible climate change action has gone.

As we leave Moree in a tiny eight-seater plane to Brisbane I look at the farmland below. I feel a huge sense of responsibility and fear for its future. We fly higher and it gets smaller and smaller until it's completely obscured by clouds.

Chapter Three

WHEN WORLDS COLLIDE

IT'S STRANGE TO FIND MYSELF EMBARKING ON A JOURNEY AROUND the world with one of the most powerful political operators in Australia. 'The Mullah Omah of the Libs,' is how Nick Minchin is described by one prominent Liberal Party moderate. But the question the politicos have been asking is whether Nick's power will survive in retirement. Says the moderate, 'It will just give him more time to devote to the game.'

From what I've seen of Nick Minchin so far, this prediction seems accurate. He is often on his mobile phone, taking and making calls about Liberal Party business. He confidently dishes out advice to the people on the other end of the line. I'm sure this advice is strategically spot on; it comes from a lifetime of experience. Nick entered politics straight after university and rose quickly through the ranks of the Liberal Party. As a 24-year-old he became a staffer for the party's federal secretariat. At 30 he was promoted to deputy federal director. Two years later he became state secretary of the South Australian party. His next stop was election as a senator, a post he held for seventeen years. Throughout this time, Nick has never hidden his deep conservatism. Journalist Paul Barry writes:

A champion of low tax, small government and the monarchy, Minchin has always been a public admirer of Enoch Powell, the brilliant but barmy British maverick of the Right, who is famous for his prediction that 'rivers of blood' would flow through Britain because the country had so many immigrants. It says something about Minchin's bravery that he would be proud to associate himself with such a man—but it also says a lot about his politics.

Like many conservatives, Minchin doesn't believe in man-made global warming, despite an army of scientists who do. He joked in his Senate valedictory that he would start up a group called 'Friends of Carbon Dioxide', whose slogan would be 'CO2 is not pollution'. His inoculation runs deep. Back in the 1990s he was publicly sceptical of the dangers of tobacco, issuing a dissenting report from a Senate committee, in which he disputed the idea that passive smoking was dangerous and that nicotine was addictive. He also claimed that the tobacco industry was over regulated. Such extreme views certainly put him right out there with Enoch.

I wonder about the source of Nick's conservative views. Perhaps he resists the science of climate change because he vehemently disagrees with the politics of many prominent advocates of climate action, such as Al Gore, a Democrat, and Bob Brown, a Green. Despite the long list of high-profile conservatives who have gone on record as accepting the science—media mogul Rupert Murdoch; US Republicans Arnold Schwarzenegger, John McCain and Newt Gingrich; British Prime Minister David Cameron and former leader Margaret Thatcher to name a few—climate change is still often seen as a left-wing issue.

This is largely due to a group of US Republican Party tacticians, who in the 1990s decided to link acceptance of human-induced climate change with left-wing beliefs. As Clive Hamilton outlines in his book *Requiem for a Species,* their goal was to characterise science as ideology. This strategy has certainly worked for Nick. He sees government action to reduce emissions as an unacceptable interference in the market. He's opposed to economic incentives for business to clean up their act, against government-mandated limits on emissions and loathes the idea of a price on carbon. The one exception to Nick's broad free-market ideology is in the area of nuclear power. He champions it consistently despite its need for significant taxpayer subsidies and government underwriting of the enormous insurance risk.

·

A 2011 study from Michigan State University entitled 'Cool Dudes: The Denial of Climate Change Among Conservative White Males in the United States' reported that 29.6 per cent of conservative white men believe global warming will never have much of an effect. This compared with only 7.4 per cent of the general adult population and 14.9 per cent of conservative white females. Polls show the breakdown is similar in Australia. When report authors Professor Aaron McCright and Riley Dunlap examined '*confident* conservative white males'—those who self-report having a high understanding of global warming—the findings were even more striking. Of these men, 48.4 per cent believe global warming won't happen versus 8.6 of other adults. To explain this 'white male effect' the researchers suggested that because, historically, white males have faced fewer obstacles in life, they are less risk-adverse.

Many psychological studies show that men are more willing to take risks. A 2010 report by the American Psychological Association

(APA) states, 'Numerous studies report differences in risk perception between men and women, with women judging health, safety, and recreational risks and also risks in the financial and ethical domain to be larger and more problematic than men.'

The researchers, McCright and Dunlap, also note another possible explanation for the lower acceptance of mainstream climate science by some men. The solutions to climate change challenge the status quo, so some conservative white males find their 'identity as an in-group' threatened. In other words, some climate sceptics who benefit from the way the economy is currently structured might be trying to protect their privileged position. They're the small proportion of the population that might benefit from an economic system that allows continued emissions of carbon pollution. They're also more willing to take the risks of ignoring climate science because they know they might be able to buy their way out of the worst climate change impacts. These factors make it easier for them to reject the science.

The report authors ultimately conclude that the study reveals more about politics than gender or class. 'When you start talking about climate change and the need for major changes, carbon taxes and lifestyle changes, [conservatives] see this as a threat to capitalism and future prosperity,' said McCright in an October 2011 article in *Scientific American* by Julia Piper and ClimateWire. 'So conservatives tend to be very negative towards climate change.'

•

Not that I think the issue of climate change is simply a political football to Nick. In 2006 and 2007 the Australian public had been in almost unanimous agreement about the science of human-induced climate change. Even then, Nick didn't waver from his conservative stance. He's not like current opposition leader Tony Abbott, who seems to encourage the sceptics when he's with them but distance himself from their views in other company. Nick wears his scepticism proudly and openly.

Cognitive scientist Stephan Lewandowsky from the University of Western Australia wrote in an 11 November 2011 article on *The Conversation* website:

> Worldviews come in many shades and forms, but one prominent distinction—popularised by Professor Dan Kahan at Yale University—is between people whose worldview is 'hierarchical-individualistic' and those whose worldview is 'egalitarian-communitarian' . . . The distinction is extremely powerful and permits prediction of people's attitudes towards numerous scientific issues.
>
> Perhaps not surprisingly, hierarchical-individualistic individuals are more likely to resist acceptance of climate science than egalitarian-communitarian individuals. Why? Because implicit in the message we get from climate science is the need to alter the way we currently do business. The spectre of regulation looms large, and so does the (imaginary) World Government or other interventions—such as multilateral agreements—that are anathema to the notion that individuals, not governments or societies, determine their own fate.

This distinction rings true for me when I think about my own worldview. I believe our world should be more equal. I believe that we have a responsibility not just to our immediate family but also to our broader community. This extends to people we will never meet, including future generations.

On the other hand, Nick certainly seems to have a 'hierarchical-individualistic' worldview. He's against compulsory voting. He acknowledges that not wearing a helmet on a bicycle is dangerous, but he thinks it should be up to the individual to decide whether or not they want to do it (above a certain minimum age), ignoring the fact that it's the rest of us who pick up the tab when people end up in hospital or permanently injured.

In Nick's speech as he left Parliament in June 2011 he'd said:

> *I entered this place with a profound commitment to smaller, less intrusive government and lower taxes, only to watch the reach of government into our lives, and the imposts upon us to pay for it, continue to expand ... I failed to achieve the sale of a government owned private health insurance company called Medibank Private. I dare not even mention what else I would like to have sold.*

Nick has been a major barrier to climate change action in Australia over the past decade. He played a key part in orchestrating the overthrow of Malcolm Turnbull at the end of 2009 as leader of the Liberal Party. As a senior figure, he's given other sceptics in the party the legitimacy to argue loudly against climate science. He bears a lot of responsibility for destroying the bipartisan support Australia used to have on climate change science when Malcolm Turnbull was leader. For all of these reasons I'm finding it hard to adjust to being around him.

I suspect that Nick is also finding it hard to adjust to being around me. Without sounding too Pollyanna-like, I've been trying to change things for what I believe is the better for a large chunk of my life. Obviously there have always been obstacles and barriers, but I'd usually focused on the things that *were* working. I'm not claiming to be perfect—far from it. I'm too bossy, sometimes selfish, rarely patient, and usually very grumpy in the mornings. I'm terrible at taking other people's advice, even when I suspect they're right. When it comes to my activism some people find me a bit too much—too relentlessly positive. Others see me as naive. I know not everyone shares my belief that a global movement of young climate advocates can change the world. So it must be as strange for Nick to be around me as it is for me to be around him.

I've mostly stayed out of head to head debates with those actively trying to discredit climate science. I think I'm more likely to make a difference by explaining the science on its own terms as simply and clearly as possible. I don't need to answer every single contradictory point raised by the professional sceptics in order to do this. I'm also an emotional sponge with a tendency to absorb others' feelings. This makes being around hard-core sceptics and climate deniers difficult and depressing. Just one day into the trip I can feel my positive energy, which I usually carry around in bucket-loads, battling with feelings of despair that Nick's arguments are getting traction in the community.

I suspect that for my friends who aren't active in climate change campaigning, their inaction is not because they don't understand or don't care. Instead it stems from doubts that there's anything they can do to change the course the world is heading in. It's the 'small cog in a big wheel' syndrome. People focus on the wheel and how big it is without remembering that even a little cog, when connected to other cogs, has the ability to change the wheel's direction.

Ever since I was a toddler I've understood that my actions could change things—not just in my own life, but also in the lives of others. I'd watched my mum successfully campaign to get the council to plant trees in our suburb. I'd seen our neighbourhood association win a battle to get our street closed off to traffic because of the large number of children in the area. My dad was an Amnesty International member. I'd seen him write letters to governments all over the world on behalf of political prisoners, some of whom were eventually freed.

My own first forays into trying to change the world had occurred in primary school. After seeing the movie *Free Willy* I found out that Keiko (the whale who played 'Willy') was being held in overheated, cramped conditions in Mexico City, where he'd been sold to an amusement park. By the time I'd seen the film, Keiko had lost weight, developed skin lesions and was expressing signs of acute loneliness. I joined the Free Willy-Keiko Foundation and set about raising money from odd jobs around the neighbourhood. My goal was to help pay for a whale sanctuary to rehabilitate and eventually release Keiko and other captive whales.

I wasn't the only one raising money: thousands of schoolchildren like me around the world donated their pocket money, too. The funds helped Keiko to be moved to a rehabilitation facility at the Oregon Coast Aquarium. He experienced natural seawater for the first time in fourteen years and gained more than 1000 pounds. His skin lesions began to heal. I still remember being excited when I heard the news that Keiko caught and ate his first live fish.

Eventually, Keiko was successfully re-introduced into the wild. When he died in 2003 he was the second oldest orca male ever to have been in captivity. Keiko's memorial service was attended by over 700 people. The veterinary chaplain said at the service, 'Keiko was not one of our kind but nonetheless, was still one of us.'

Now, some of you reading this might be shaking your heads at all this saccharine whale-saving sweetness. I get that. But being part of Keiko's journey was a turning point for me. It taught me that people *could* make a difference to others' lives—whether whale, or human. I took this newfound understanding to heart. I joined my dad in writing Amnesty International letters. I turned off lights to save energy. I helped my elderly neighbour walk her dog. Basically, I wanted to try to help everybody.

When I was 14 years old, I started volunteering with the Wilderness Society after their campaign director gave a speech at my school assembly. I assisted with their campaigns to protect the Hunter Valley's forests, and then decided to start an environment group at my school. We called ourselves the Merewether Greenies. Attendance wasn't a problem, because I bribed friends and total strangers into coming with the lure of free Tim Tams. We convinced my principal to make a new official school sport called 'environmental activities'. I ran it. It consisted of setting up recycling systems, planting an area of the school with native plants and grasses, and writing passionate handwritten letters to politicians asking them to save the environment. We campaigned against the proposed Jabiluka uranium mine, raised money for the Cancer Council and collected signatures against animal cruelty.

But soon we were ready for something bigger. Our next step was quite a leap—we took on the New South Wales state government and mining giant BHP Billiton as part of a campaign to protect an amazing place north of Newcastle called Stockton Bight, which was under threat from sand mining. Stockton Bight is a 32-kilometre stretch of the last great sand dune and paperbark forest system in New South Wales. It is also brimming with Aboriginal sacred sites and endangered species.

The campaign swallowed up hours of my weekends and my time after school. We participated in meetings between our coalition of community groups and the Traditional Owners. We dropped banners from the highest building in Newcastle (the council car park). We harangued people at music festivals to collect signatures on petitions. We dressed up as koalas and bushrangers to get in the local media.

Finally all of this activity, and the hard work of the other groups in the coalition, paid off. In my final year of high school, on 21 February 2001, the NSW government declared a large part of Stockton Bight a conservation area to be managed by the Traditional Owners under the *National Parks and Wildlife Act*. It wasn't a perfect victory—parts of the beach are still being mined. But it was a huge achievement.

The whole experience taught me a fundamental truth that I carry with me to this day. I learned that in Australia change comes not from the top down, but from the bottom up. Change can happen even when you're up against one of the world's biggest mining companies. Change can happen in spite of a pro-developer state government. Change can come when people join together and don't give up until they've made a difference.

It might sound strange, talking so much about events that happened in high school. But these experiences laid the foundation for the work I did later. They made it easy for me to throw myself wholeheartedly into student environmental activism during my first year of university. Then they gave me the confidence to drop out of university for a year in 2005 when I was elected National Environment Officer for the National Union of Students. In that year I moved to Melbourne to coordinate a nationwide university clean energy campaign. I travelled from campus to campus, training student organisers and negotiating with university administrations to win campaigns for green power and energy efficiency programs.

At the end of 2005 the Canadian government invited me to travel to Montreal for the United Nations climate talks. There I met young climate activists from around the world. They were so awesome that I decided I needed to do more in Australia. I resolved to set up a coalition of youth organisations and young people. Our mission would be to build a generation-wide movement to solve climate change before it was too late. The Australian Youth Climate Coalition (AYCC) was born.

It sounds like the end of the story when I write it like that, but it was really just the beginning. I was 22 years old when I had the idea to set up the youth climate coalition. I then spent the next five years of my life living and breathing that dream with my co-director Amanda McKenzie and our team of incredible young organisers.

Today, the AYCC is one of the largest youth-run organisations in Australia. We have over 71,000 members, thousands of active volunteers and six full-time staff under the age of 26. We've been able to move young people from being just a demographic into a constituency who will make choices—both consumer and political—on the basis of climate change. We've helped reframe climate change as an issue of intergenerational equity and have reminded a whole lot of politicians, business leaders and everyday Australians that it's our future at stake in this debate.

I left the AYCC as a staff member when I turned the ripe old age of 27, the age limit I had set for myself when I established it. I moved onto the board as a hands-off adviser and took up a new role at a small communications agency for non-profits founded by a group of former AYCC and GetUp staff. Our mission is 'to help people do good, better'.

So that's my story in a nutshell. I've been working on climate change on and off for half my life. Sometimes it's exhausting, sometimes it's amazing, sometimes it's depressing, and always it feels like an uphill battle. I live in hope that maybe the world will wake up in time and act with sufficient bravery to avoid the worst impacts of climate change. This is what I am working towards in this journey with Nick Minchin.

Chapter Four

PILGRIMAGE TO PERTH

WE'RE ON A PILGRIMAGE TO THE SPIRITUAL HOME OF AUSTRALIAN climate denial. Our destination is a house on the outskirts of Perth belonging to blogger Joanne Codling (although she also uses the stage name Jo Nova) and her husband David Evans. They are Nick's first choice to try to change my mind on the science of climate change.

I spend the long plane trip to Perth reading as much climate science as I can cram into my brain. My key references are reports by the Australian Academy of Science and the United Nations Intergovernmental Panel on Climate Change (IPCC). As I read, the back of my mind is pettily stewing over the fact that Nick's in business class and I'm stuck in economy. The reading is far more useful than the stewing. The whirr of the engines, the lack of distracting influences such as Facebook and the cramped space mean that I soak up a surprising amount of information on the five and a half hour flight. I've read plenty of popular climate science texts before, but now the presence of Nick and a television crew, and the pressure to perform in front of people who could be quite hostile, adds an edge to my frantic cramming. It's easy to be overwhelmed by it all but ultimately I'm reminded that climate change science comes down to three straightforward facts.

The Australian Academy of Science set out these facts in its 2010 document *The Science of Climate Change: Questions and Answers*. The first fact is that humans have, since the Industrial Revolution, sharply increased the concentration of greenhouse gases in the atmosphere. The second is that these greenhouse gases have a warming effect. It's a basic principle of chemistry that greenhouse gases trap heat. This was first discovered by British naturalist John Tyndall in his laboratory experiments in 1859 and confirmed by Swedish scientist Svante Arrhenius in 1896. The third fact is that global average temperatures have increased dramatically since the Industrial Revolution. The laboratory experiments have moved into the real world: temperature records for our planet's air, oceans and surface show an unequivocal warming trend. The last decade has been the hottest since temperature records began.

These three facts get us to a point where scientists agree that the planet is warming and the cause is human emissions of heat-trapping greenhouse gases, which I also refer to as 'carbon pollution'. 'Warming of the climate system is unequivocal,' stated IPCC's 2007 report, based on the work of over 2500 contributing scientists. It is 'very likely' (defined as more than ninety per cent certain) that human activities such as the burning of fossil fuels account for most of the warming in the past fifty years. The probable temperature rise by the end of the century will be between 1.8 and 4 degrees Celsius. The impacts will be severe. They will likely include more extremely hot days, more extremely cold days and more flooding. Arctic summer ice is likely to disappear in the second half of the century. Parts of the world will experience increases in the number of heatwaves and the intensity of tropical storms. This has flow-on effects for agriculture, infrastructure and human health.

Climate change isn't caused by CO_2 alone. There are other greenhouse gases that also kick in. Carbon dioxide, however, lasts in the atmosphere for centuries so it's arguably the most significant. The level of warming from a doubling of carbon dioxide concentrations *alone* over pre-industrial levels will be 1.2 degrees Celsius globally. Scientists are fairly united about that number. Even Nick's spokesperson, Joanne Codling, doesn't dispute it when pressed, as I'll soon find out.

Climate change isn't just about future changes. Already, with just a thirty-eight per cent increase in carbon dioxide, the world has warmed by 0.75 degrees Celsius. This is the *global average* temperature increase, which means that some places have warmed less and some more. The closer a place to the North or South poles, the warmer it has become. The Arctic, for example, has already warmed by around 6 degrees Celsius over pre-industrial temperatures. This has led to glaciers collapsing into the ocean, dramatic images of which we see on television accompanying stories about climate change. Climate change has also already caused more extreme weather events worldwide. We've seen longer and harsher droughts, more intense cyclones and more of the hot, dry conditions that foster mega-fires and cause food crops to fail. All of these things are consistent with predictions from the climate models used by scientists. According to the global insurer Munich Re, climate-related disasters in 2010 caused US$130 billion in losses, rising food prices and 21,000 deaths in the first three-quarters of the year alone. If all this has happened from just a thirty-eight per cent increase in CO_2, imagine what will happen with 100 per cent increase. The scary thing is that this is the path we're on unless the world wakes up and makes big changes before it's too late.

On top of this, we have to remember that the earth's climate system is not linear. In addition to the simple equation that a doubling in CO_2 leads to 1.2 degrees warming, positive feedback mechanisms cause the world to warm further. This is how the IPCC gets its prediction of global average warming of between 1.8 and 4 degrees. Water vapour is a good example of a feedback mechanism. Carbon pollution leads to temperature increases, which means warmer air. Warm air means more evaporation. More evaporation means more water vapour in the air. Water vapour is also a greenhouse gas, meaning it holds more heat. This leads to even higher temperatures.

There are many feedback mechanisms at play in the earth's systems. Consequently, when we talk about how much additional warming we expect over and above the initial 1.2 degrees from a doubling of carbon pollution alone, we're talking about a range. The reason it's a range rather than a definite single number is because of what scientists call 'climate sensitivity'. The big question is this: for every one degree of warming from carbon dioxide, what will the total amount of warming be? Some scientists, such as NASA's chief climatologist James Hansen, are fairly sure that the climate sensitivity is around 3. That is, for every one degree of warming from CO_2, there's 3 degrees of warming in total. However, this isn't yet completely agreed upon in the scientific community. It's this room for doubt that seems to fuel climate scepticism and leads to the statement Nick loves to repeat: 'The science isn't settled.' It's true that the science around feedback mechanisms isn't completely sorted out. But it's just not true to say that the basic science of the greenhouse effect and human-caused climate change isn't settled. It has been for a long time, and it's misleading to say that it hasn't.

As if feedback mechanisms weren't scary enough by themselves, the climate system has certain tipping points. Tipping points are thresholds that, when passed, will change the climate system dramatically and irrevocably. They are the points of no return. There's no simple equation for the level of carbon pollution that will trigger temperatures that get us past tipping points. Like a line of fishing wire across a footpath, you don't see the booby trap until you've tripped over it.

Tipping points are critical to any discussion of climate change. James Hansen devotes a fair chunk of his 2009 book *Storms of My Grandchildren* to these points of no return. He explains that the urgency of action on climate change 'derives from the nearness of climate tipping points, beyond which climate dynamics can cause rapid changes out of humanity's control'. He describes how tipping points occur because of feedback mechanisms and uses the example of a microphone and speaker. We've all been at an event where the microphone is too close to the speakers and picks up the amplification, which is again picked up by the microphone, until the noise has everyone in the crowd covering their ears. Dr Hansen's fear is that we're on a path to handing our children a world where tipping points have been passed. This is because of the way the original warming (the original sound) is amplified by feedback mechanisms (the sound bouncing off the speaker). In Hansen's words this means bequeathing our children 'a dynamic situation that is out of their control'. It would mean the microphone had been super-glued to the speaker and couldn't be moved.

So what are the feedback mechanisms that will lead to these tipping points? I mentioned heat-trapping water vapour briefly but sadly there are many more. Journalist Jo Chandler in her 2011 book

Feeling the Heat writes that tipping points include: 'the melting of ice sheets as warm waters erode them from below; the bubbling up of methane hydrates, releasing potent greenhouse gases when warming opens their crypts under the sea floor; the disabling of the powerhouse rhythms of winds, rains and currents on which nature turns and civilisation is founded.'

Let's take one of these examples—the melting of ice sheets—and look at how it works in practice. We already know that greenhouse gases trap heat. This means that when humans emit more carbon pollution, more heat is trapped in our atmosphere. It can't escape back into space like it used to be able to do before humans emitted record levels of carbon pollution. A hotter atmosphere means higher air, land and ocean temperatures. In fact, about ninety per cent of global warming is trapped in the oceans. So when you have a combination of warmer waters and warmer air, the Arctic starts to melt more quickly. When it melts it turns from white ice to dark blue—almost black—ocean. We all know to wear white rather than black clothes on a hot day. Why? Because white reflects heat and black absorbs it. What's the lesson here for the climate? We used to have a large surface area of white ice reflecting heat. Now we have a large surface area of dark ocean absorbing heat. As more heat is absorbed the oceans warm even more quickly. This increases the melting rate not just of other ice but also the permafrost.

The permafrost is a big frozen mass of greenhouse gases including carbon dioxide and methane, trapped among frozen soil, rocks and organic matter. Much of it lies under Siberia and around the Arctic Circle. When it melts it starts to release the methane it's storing back into the atmosphere. It melts not from the sun shining on it from above, but from the warm waters lapping at it from below. A

2011 report from the World Meteorological Organization says that methane levels are rising from their formerly stable state and this could be a result of the permafrost starting to destabilise. 'Now more than ever before we need to understand the complex, and sometimes unexpected interactions between greenhouse gases in the atmosphere, Earth's biosphere and oceans,' said WMO Secretary-General Michel Jarraud.

A 2009 report from the US National Academy of Sciences lays out the situation, noting that current emissions of greenhouse gases from humans 'have already committed the planet to an increase in average surface temperature by the end of the century that may be above the critical threshold for tipping elements of the climate system into abrupt change with potentially irreversible and unmanageable consequences'. They say that this means the climate system 'is close to entering if not already within the zone' of dangerous anthropogenic (human-induced) interference.

When will humans have emitted so much carbon pollution that we pass earth's tipping points? We don't know for sure. And one doesn't normally think of the climate system as secretive or sneaky. (After all, Mother Nature has shown herself to be big, bold and capable of inflicting serious, in-your-face damage.) But as NASA's James Hansen explains in *Storms of My Grandchildren*, climate change's real danger is that 'it can sneak up on us. By the time people recognise that big changes are under way and begin to take action, the system may already have enough momentum that it will be very difficult, if not impossible, to prevent catastrophic effects.' Hansen believes we passed tipping points when we reached 350 parts per million of CO_2 in the atmosphere. (We're currently at 392, so he argues we must stabilise carbon pollution and then start taking carbon out of the atmosphere). At 350 parts per million, we're looking at a global average temperature rise of 1.5 degrees Celsius over pre-industrial levels. (Remember, we've already had a 0.75 degrees Celsius increase.) Other scientists believe we'll pass tipping points at 450 parts per million of CO_2 in the atmosphere. This would translate into a global average temperature rise of 2 degrees Celsius.

The next five years are crucial if we are to avoid runaway climate change. As 2011 drew to a close, the usually conservative International Energy Agency took the unexpected step of going public with a deadline for avoiding dangerous climate change. The IEA chief economist, Dr Fatih Birol, warned that if the world continues on its present pathway, all the allowable emissions for stabilising climate change at 2 degrees will be locked in by existing infrastructure in just five years. The world has already emitted about 80 per cent of the emissions 'budget' that would be required to stay under the 2 degree trajectory. 'If, as of 2017, we do not see a major way of clean efficient technologies then the door to 2 degrees will be closed and will be closed forever,' stated Dr Birol. Anything built from now on that produces carbon will do so for decades, and lock in irreversible climate change.

The reason that the next five years will determine our future is that CO_2 persists in the atmosphere for centuries. Scientists Kevin Anderson and Alice Bows from the Tyndall Centre for Climate Change Research have crunched some important numbers. They found that even if global emissions peak in 2020 and fall by 3 per cent globally after that (a set of very optimistic assumptions given the current state of the international climate negotiations) it still wouldn't be enough to stabilise carbon dioxide levels at 450 parts per million, the level often associated with global warming of 2 degrees Celsius.

It would instead see concentrations rise to 650 parts per million of carbon. This level translates into a global average temperature rise of 4 degrees.

This is well above the temperature associated with the tipping points I just mentioned.

•

After arriving in Perth, the crew films some setup shots of Nick and me in a park overlooking the city. As soon as we drive in the park entrance, I recognise it. Funnily enough it's the same park where I'd helped to illegally hang an enormous banner off a suspension bridge at sunrise one morning six years ago. The banner had read 'Ignore it and it will go away' and featured a picture of earth from space. Our hope was that thousands of Perth residents would see it as they commuted to work. We wanted the banner to prompt them to think about the environment in the lead up to the West Australian state election. I'd been in Perth helping student environment groups in my role as National Environment Officer for the National Union of Students. At the time, we furtively roamed Perth at night, climbing up ladders to paste speech bubbles with environmental messages onto billboard ads. Then, one night a police officer pulled over our car and accused us of having a ladder to break into someone's house. We sheepishly pointed at the billboard across the road. No, we'd said, we weren't breaking and entering. Just pasting environmental messages onto billboards. The confused police officer let us go.

Six years later we're having trouble with the Perth authorities again. The ABC hadn't applied for a permit to film in this particular park. The rangers have noticed the camera gear and point at us with concerned looks on their faces. Max finishes his shots of the car driving around the park and we get out of there to begin the long drive out to Joanne and David's house on the outskirts of Perth. Nick reads out an introduction to Joanne and David as we drive. He's been corresponding with them via email but hasn't met them in person yet.

Joanne Codling and David Evans are a strange choice for Nick. From what I can find, neither has any formal qualification in climate science. Joanne studied molecular biology as an undergraduate before completing a graduate certificate in science communication and working as an associate lecturer. She then jointly coordinated the Shell Questacon Science Circus and wrote a children's book called *Serious Science Party Tricks*. At some point in her career she became a professional climate sceptic, despite having no specific climate change qualifications. She started a blog on the topic, and began doing speaking gigs.

Most recently Joanne has written two booklets for climate sceptics. The first is sixteen pages long and called *The Skeptics Handbook*. The second, twenty pages long, is even more polemical: *Global Bullies Want Your Money*. They're both glossy, full-colour affairs, sprinkled liberally with cartoons drawn by Joanne herself. There's a hand-drawn picture on the front of *Global Bullies* of an old-fashioned scale, like the kind used to weigh gold. The scale is labelled 'carbon trading' and features two baskets: 'risks' on one side and 'benefits' on the other. Sitting in the 'risks of carbon trading' basket are a collection of people: a suit-wearing businessman shooting himself in the head, an unemployed beggar, a starving African child, and a wedding photo of a bride and groom being ripped in half. The 'benefits of carbon trading' basket is empty apart from an overweight man clinging to the bottom covered in money. The drawing seems to insinuate the carbon trading will lead to suicides, starving African children, wide-scale unemployment and . . . the destruction of the institution of marriage. Inside *Global Bullies* Joanne rails against 'big government', scientists ('science has been corrupted'), banks ('sub-prime carbon is coming') and the role of consensus in science ('there is no consensus, there never was, and it wouldn't prove anything even if there had been').

Joanne's husband David Evans has a PhD in electrical engineering. With two undergraduate and three masters degrees, he's obviously a smart guy and an expert in mathematics. But that doesn't mean he's smart about *everything*. Like his wife, he's never written a peer-reviewed article on climate change. From 1999 to 2005 he worked for the Australian Greenhouse Office as a computer modeller. He designed a carbon accounting system that was used by the Australian government to calculate how much carbon is used and released through various land management practices. Since then, he has joined his wife in her crusade against climate science.

Joanne is associated with a US conservative think tank called the Heartland Institute, which is partly funded by fossil fuel companies. The Heartland Institute partly paid for her *Skeptics Handbook* to be printed and for *Global Bullies Want Your Money* to be written. David's writings also appear on the institute's website. The Heartland Institute received hundreds of thousands of dollars from Exxon Mobil between 1997 and 2005, before a public outcry (including pressure from the UK's Royal Society) forced Exxon Mobil to stop donating. Heartland is famous for organising regular conferences of climate sceptics. The journalist Naomi Klein described one such conference she attended as:

the premier gathering for those dedicated to denying the overwhelming scientific consensus that human activity is warming the planet . . . Over the course of this two-day conference, I will learn that Obama's campaign promise to support locally owned biofuels refineries was really about 'green communitarianism,' akin to the 'Maoist' scheme to put 'a pig iron furnace in everybody's backyard' (the Cato Institute's Patrick Michaels). That climate change is 'a stalking horse for National Socialism' (former Republican senator and retired astronaut Harrison Schmitt). And that environmentalists are like Aztec priests, sacrificing countless people to appease the gods and change the weather (Marc Morano, editor of the denialists' go-to website, ClimateDepot.com).

It seems evident that companies like Exxon Mobil would try to protect their profits by delaying a low-carbon future, so their funding of organisations like the Heartland Institute would affect these organisations' objectivity. However, I doubt that Joanne and David are on their anti-climate science mission for the money. It seems it's much more about their libertarian, anti-government ideology. Joanne seems to argue that 'a privileged few' are about to usher in an age of hyperinflation in order to cement their hold on power. 'You might think inflation and climate science are only linked metaphorically,' she writes on her blog. 'But the corruption in science is fed by the corruption in our currencies . . . If the system is swimming with easy money, people can "afford" to build wildly extravagant and unproductive things, like wind-farms, carpets of solar panels, or symbolic rivers of blue plastic.' Her explanation for this looming problem? 'The monetary system that allows a privileged few to print money is the same that allows massively misdirected spending.'

•

Even though Joanne and David might have different, sometimes seemingly bizarre, motivations to their backers, it is still important to know what each stakeholder's motivations are. As Nick and I are driven out to Joanne and David's home, I'm buoyed by the knowledge that I have expert, objective, scientific consensus on my side and a sound knowledge of basic climate science facts. They would have to make incredibly strong arguments to be able to prove beyond reasonable doubt that human-induced climate change isn't happening. I'm looking forward, in a way, to seeing how they're going to try to do this.

Chapter Five

KITCHEN TABLE SCIENCE

AS WE PULL UP OUTSIDE JOANNE AND DAVID'S HOUSE, A BEARDED MAN in a khaki shirt appears suddenly in front of our car. He's holding a home video camcorder and it's trained squarely on me. 'Max? Who is that guy? ' I ask nervously. 'Is he with the ABC?' Both the camera and the way he's holding it suggest he's an amateur videographer.

It is mid morning and this turn of events feels quite dramatic for the quiet neighbourhood where Joanne Codling and David Evans live. It's the kind of suburb where the benefits of affordable land and child-friendly streets outweigh the inconvenience of the long drive required to reach Perth's city centre. There's a tree and some children's toys in the couple's front yard and a car in the driveway. The house directly opposite bears a large sign proclaiming that it is solar-powered. (I wonder what the neighbours think of Joanne and David, and vice versa.) But I didn't have much time to observe anything else in the street before the camera-wielding man appeared.

Max gives me a look of resignation. He explains that when he had arrived at Joanne and David's house earlier that day, the couple had issued an on-the-spot ultimatum. Despite previously agreeing to be filmed, Joanne and David demanded that their friend Barry be present to film the meeting or it wouldn't go ahead at all and we'd have to leave their property. This was unexpected to Max but I'm not surprised to hear it. 'Conspiracy theorists' is a common phrase used in reference to Joanne and David. And I know they don't think much of the ABC: Joanne's blog accuses the national broadcaster of being a vehicle for 'propaganda'.

Having Barry film me as I get out of the car is unsettling. I'd agreed to be filmed by the ABC, not him. Most people don't particularly like others owning footage of them. It feels creepy. And since it's completely thrown me off, it also feels like an advantage for Nick. How would Nick feel if he turned up to one of the meetings I'd organised and was suddenly ambushed by an unknown video cameraman? Nick says he didn't know about the situation or have anything to do with it. 'I'm asking you to trust them,' he says. 'I have complete faith in these people but . . . It's them just taking an insurance . . . it's their insurance policy against them being made to look like idiots.'

So I have a choice. I can refuse to be filmed by Barry and we won't do the meeting with Joanne and David. This might be a prudent decision. Nick would have to find a replacement sceptic for us to meet later on in the journey. Or I can decide to agree for two people who are opposed to everything I stand for to own footage of me that will probably be broadcast on Joanne's blog.

In the end I feel for Nick. He looks dejected by this turn of events. I've prepared for this meeting and we've come all the way to Perth. Think of all the wasted carbon emissions if we come home with no footage! And if I do say no, Joanne's blog will undoubtedly feature a story the next day along the lines of 'Climate activist afraid to be filmed'. So I say yes. Maybe I've just made a mistake: time will tell.

I walk in and shake hands with Joanne and David. My mum and grandmother taught me never to show up at someone's home empty handed, so I present them with a large punnet of ripe strawberries I'd purchased that morning.

Joanne is a petite, slender blonde and dressed conservatively. Her eyebrows are perfect and her hair is tied in a neat ponytail. She's articulate and polite. She'll come across well on television, I think. She welcomes me into her home with a smile, thanking me for being 'good-natured' about Barry being there. 'Hopefully we never even use it,' she says, referring to the footage. I have a hunch this isn't true, so Joanne and I agree that if she posts it, it will be in full rather than selectively edited. Joanne's husband David is harder to describe, and smiles less. He's broadly built, with a pale pink complexion. His hand is sweaty when I shake it. I think he's nervous.

We walk through the house and into their open-plan kitchen. I have a quick glance at their bookshelves, which are filled with science fiction novels and maths textbooks. Barry moves to one side and sets up his camera on a tripod. Max arranges us around the kitchen bench. If I'd known we'd be standing for the next two hours, I wouldn't have worn my new pair of heels.

David begins. He tells me he used to be 'out there to save the planet' and previously worked at the Australian Greenhouse Office. 'I was also a member of the Labor Party on and off for fifteen years and I was a member of Greenpeace, so I was a believer. I thought I was saving the planet and I thought this problem was a really momentous one that needed solving straight away,' he says, barely drawing a breath. 'As the years went by I found out more evidence, the evidence supporting it drifted away and evidence started accumulating that the man-made hypothesis wasn't true. And so I changed my mind.'

David says that after he finished working at the Australian Greenhouse Office 'for other reasons', he looked into climate change a little more. 'I'd noticed that the evidence wasn't right and somewhere around 2006 or so I noticed that it was quite wrong,' he says.

Joanne tells me that she also used to accept the science of climate change. But this changed, she says, when 'David comes to me one day and he says to me, he says, "You know there's no evidence left to support the great global warming scare."'

They looked at the ice core data together. Joanne describes this as a kind of awakening: the moment when a mum and dad team from Perth discovered that thousands of climate scientists and all the world's main scientific academies were wrong. 'I thought I knew stuff about this debate and this floored me because suddenly I realised that there was a lot that was going on behind the scenes that wasn't being reported,' she says.

I should point out that Joanne and David seem to accept, for the most part, that climate change is happening. There's too much undeniable direct evidence in the real world for them to argue otherwise now. Glaciers are melting, the Arctic is shrinking, natural systems are changing and the ocean is becoming more acidic. 'Anna, we agree that carbon dioxide causes global warming,' says David. 'Every molecule of carbon dioxide that we emit causes some global warming. It's not a question of *if* it causes warming or not, it's a question of how much.'

The confusion arises because Joanne and David refuse to accept climate change is caused by *human* emissions of carbon pollution. Although, sometimes they say human carbon emissions cause *some* warming—just not enough to worry about.

'Now the problem is,' says David, 'in the climate models only one third of the warming is due to that direct effect of CO_2, the [other] two thirds is due to positive feedbacks and that's the bit we're arguing about.'

Joanne and David believe that since human emissions of CO_2 will only cause a *direct* temperature rise of 1.2 degrees Celsius as a global average, something else—other than humans—must have caused the warming we're experiencing now. And that something, according to them, definitely can't be feedback mechanisms. (Mechanisms that amplify the direct warming from CO_2.)

ME: *So we're just arguing . . . we all agree that climate change is happening?*

DAVID: *Yes.*

ME: *It's due to carbon pollution.*

DAVID: No.

ME: *But we're just saying about how much . . .*

DAVID: *No, I agree that climate change is occurring but I don't think it's due to carbon dioxide emissions.*

JOANNE: *There's some small immeasurable amount.*

DAVID: *Some very small amount is but the majority is not.*

ANNA: *Potentially 1.2 degrees . . .*

DAVID: *If carbon doubles.*

JOANNE: *We have no real beef with that calculation, it's the question of these feedbacks amplifying that 1.2 to 3.5, or in the case of Garnaut* [the economist who summarised the science and impacts in a major report for the Australian government], *I think he's talking 5.9 or 5.7 degrees or something.*

ME: *So you don't believe in positive feedbacks?*

JOANNE: *We don't think the positive feedbacks is supported by any evidence.*

It's hard to pin down what we're actually arguing about, because Joanne appears to contradict herself. She tells me that there's 'some small immeasurable amount' of warming from CO_2. This was the central claim of her handbook too—that the effect of CO_2 is 'so small, it's unmeasurable'. Yet in the next breath, she and David reassure me that they actually *do* accept a doubling of CO_2 causes global warming of 1.2 degrees. This sounds pretty 'measurable' to me. 'If CO_2 doubles from pre-industrial times,' she tells me, '1.2 is the maximum [temperature rise] that there's any evidence to support. That would be the direct effects of CO_2.'

This is a big admission, because not only can a global average temperature rise of 1.2 degrees be measured, it can also cause significant damage. A global average increase of 1.2 degrees would make land masses about 2 degrees hotter (even more in the middle of continents). And it would make the higher latitudes, closer to the poles, much warmer—around 4 or 5 degrees. Even without feedback mechanisms, this level of increase has the potential to cause major disturbances in natural systems and human infrastructure.

Joanne and David take us through a pile of coloured graphs and photographs. These put forward many of the same arguments that form the basis of Joanne's *Skeptics Handbook*. Each of them has been rebutted many times by climate scientists. Even the fearless leader of the sceptical scientists, Dr Richard Lindzen, has rebutted one of Joanne's claim about a 'missing hotspot'.

Joanne and David don't agree that rising levels of carbon pollution lead to significant temperature increases. They point to graphs from ice cores drilled at Vostock, Antarctica, to argue that because CO_2 wasn't the main cause of temperature increases in the ancient past, it can't be the cause today. These graphs show that when CO_2 went up hundreds of thousands of years ago, temperatures didn't follow straight away. Back then there was an 800-year lag between the rise in CO_2 and the temperature increase. Joanne and David argue that because of this lag, it wasn't the CO_2 increases alone that caused the warming back then. And they're right. But this doesn't mean it isn't CO_2 that's causing the warming our climate is experiencing *today*.

Figure 5.1 shows the graph we're talking about, constructed from Antarctica's ice core records and published in a peer reviewed paper in *Nature*. The ice cores are long cylinders carefully extracted deep from the belly of the ice. When the core is pulled out, it brings with it layers from 400,000 years ago. These ice cores are goldmines of information about the earth's past climate because of the air bubbles trapped safely inside them. These bubbles are analysed in laboratories to show us the composition of the air, way back through ancient history.

Figure 5.1 shows a clear correlation between rising CO_2 levels and rising temperatures. But as Joanne and David correctly point out, the rise in CO_2 comes *after* the rise in temperature. 'Something else caused the warming!' scream large letters in Joanne's *Skeptics Handbook*.

Joanne is right: ancient changes in climate were triggered by natural things like the earth wobbling in its orbit and getting closer to the sun, changes in the intensity of the sun's rays and mega-volcano eruptions. In climate-scientist speak all these things are called 'forcings'—which means 'something with the ability to change the climate system'.

In ancient history, natural forcings made the earth heat up and cool down. We saw enormous changes in temperature—big enough to send the earth tumbling in and out of ice ages—but these weren't

Figure 5.1 Temperature and CO_2 concentrations in the atmosphere over the past 400,000 years

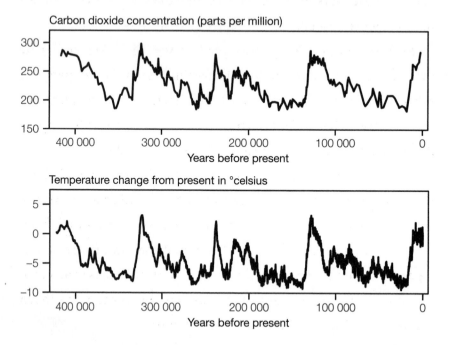

Source: Adapted from Philippe Rakacewicz, UNEP/GRID-Arendal, *Vital Climate Change Graphics*, 2000, www.grida.no/publications/vg/climate/page/3057.aspx

caused *exclusively* by the initial natural forcing. Something else happened to amplify this initial warming. In some cases this caused quite abrupt climate shifts. What the ice cores tell us about this amplification is critical to understanding the climate change we're witnessing and causing today.

In the past the natural forcing, such as the earth wobbling closer to the sun, caused temperatures to rise. These higher temperatures *then* led to higher concentrations of greenhouse gases, including CO_2, which meant the earth heated up even more. It's a simple process: when the atmosphere warms, it causes the ocean to heat up, which means more evaporation, which means more heat-trapping water vapour. Water vapour 'amplifies' the effect of CO_2. Voilà! Here we have evidence of feedback mechanisms at work in the past.

So, we know two things about the ice cores. First, Joanne is right in saying it wasn't carbon dioxide causing the initial temperature rise back then. Second, we know that once the initial temperature rise occurred, CO_2 and other greenhouse gases kicked in to amplify it through feedback mechanisms. It's a vicious cycle.

I should note that Joanne disagrees with climate scientists about the ice cores providing evidence for feedback mechanisms. 'Maybe there's a small amplification in there,' she says, 'but it hasn't been shown mathematically, it hasn't been published in peer-reviewed research.' I don't know how she can argue this, because the earth could not have moved out of the last ice age without feedback mechanisms—we'll have to agree to disagree.

But the question remains, if CO_2 didn't kick off climate changes in the past, what makes scientists so sure it is the cause now? Couldn't it be that pesky sun again, getting hotter? Or maybe the earth is wobbling again on its orbit? Or have there been a cluster of mega-volcanoes spewing greenhouse gases into the atmosphere somewhere that we haven't noticed?

The answers to those questions are no, no and no. These days we have tools to measure the earth's orbit, the intensity of the sun and the activity of volcanoes. Scientists have proven unequivocally that the earth is *not* moving closer to the sun, the sun's rays aren't getting any stronger and volcanoes have only a minor, temporary influence. The Australian Academy of Science concludes that 'all the trends in the Sun that could have had an influence on the Earth's climate have been in the opposite direction to that required to explain the observed rise in global average temperatures'.

Sadly, it's no longer natural forces driving most of the changes in our climate system. There's been a new forcing on the block since the Industrial Revolution and its name is humans. Or to be more precise, human emissions of greenhouse gases from burning fossil fuels and clearing forests. It's easy to record human emissions of carbon pollution—they're measured at our next destination in Hawaii as well as at thousands of other locations around the world. In addition, countries publish their emissions data each year. So we know we've got increased amounts of heat-trapping greenhouse gases like CO_2.

But just in case all this isn't enough evidence that significant climate change is being caused by humans, there's one additional, compelling piece of proof. Satellites in space have, for the past few decades, measured how much heat is going into the earth's atmosphere and how much is coming out. Satellite records now show that roughly the same amount of heat from the sun has been streaming in, at the same rate, over the last few decades. However, the amount of heat coming out has dropped.

So we have the same amount of heat from the sun going in, but since large-scale carbon pollution started, less heat coming out. Scientists call this an energy imbalance. If the heat's not coming out, it's being trapped by something—those pesky heat-trapping greenhouse gases, floating around in our atmosphere as a result of human pollution. This is called the greenhouse effect and it's a basic and observed principle of science. The gases stop extra heat escaping into space where it's supposed to go. Instead the heat gets stored in our air, land and oceans.

This is all very well established, so I figure Joanne and David must have a pretty extraordinary explanation if they're going to prove it's something other than CO_2 causing the warming. I try to coax Joanne into telling me what's leading to climate change, if not carbon pollution.

ME: *You think it's the sun that causes climate change?*

JOANNE: *We think that there's a lot of other factors that need to be considered.*

ME: *Climate scientists have considered the sun [and] they've considered orbital changes.*

JOANNE: *The IPCC [Intergovernmental Panel on Climate Change] does not include any solar magnetic affects in any of its models. The empirical evidence suggests that there's a connection between sun spots and the climate but we're not saying—we don't know what causes climate change. We're simply saying there's a lot of other factors and the models ignore them all and then say, yes, well we looked at everything and we couldn't explain the warming except for CO_2 in our models. It's argument from ignorance; it's a logical fallacy.*

Despite telling me she doesn't know, Joanne devoted a section of her *Global Bullies* booklet to sunspots as the possible cause of climate change. Under the heading 'If Carbon Didn't Warm Us, What Did?' she writes that 'people have known for 200 years that there's some link between sunspots and our climate'.

We're going around in circles by now, so I point out that the argument we're having is hypothetical. The fact remains the world *has* actually warmed since the Industrial Revolution, when humans started to emit CO_2 on a large scale by burning fossil fuels. Surely Joanne can't argue it's a mere coincidence that humans have pumped just the right amount of heat-trapping carbon pollution into the atmosphere to account for the amount of warming we've seen?

'Surface temperatures have increased and ocean temperatures have increased,' I say. I can't see how Joanne and David can dispute this, given that at the start of our conversation they accepted that climate change is happening.

'According to the dodgy thermometers, yes,' replies David.

Joanne and David now seem to backtrack. They're now claiming that warming may *not* be taking place to the extent that the temperature records show. They argue that too many of the thousands of thermometers placed around the world to measure surface temperature are placed near car parks, airport runways and heat-emitting air conditioners. Jo's *Skeptics Handbook* proclaims that 'the main "cause" of global warming is air conditioners'. It includes photos of temperature measurement stations that are indeed placed next to air conditioners.

I sigh. This argument comes from a sceptic blogger called Anthony Watts. He took photos of temperature stations in locations close to urban heat sources and developed a belief that the temperature record was wrong. I've asked climate scientists about this. They told me, yes, they know about the issue, and therefore they exclude potentially problematic data from the temperature record used to track the pace of global warming. The Australian Bureau of Meteorology *already accounts for* any problems caused in the data from those weather stations that so outrage Joanne and David. If a station is affected by urban heat, scientists simply remove them from their data sets.

'This one's in Australia,' says David. He points at a measurement station in Sydney City that the Bureau of Meteorology's website shows is excluded from annual temperature analyses because of its proximity to urban heat.

'This is the Sydney Observatory, that's the Western Distributor there, that's where the thermometer is.'

'But this is not included in the temperature records,' I say, referring to the nationwide data sets used to measure the pace of global warming.

'Sure it—' says David.

'It's not,' I interrupt. 'I mean, you've been told this before.' I'm aware of a correction by a Bureau of Meteorology staff member on one of David's web posts, clarifying this point, over a year ago. 'Why don't we call the Bureau of Meteorology now, if you want to go over it again? Let's do it, let's call them.'

Nick interjects: 'What are you saying, that that thermometer, measurements from that, is not used?'

'I've never been told anything by the Bureau of Meteorology,' says David.

'OK, well it was my understanding that this had been pointed out to you by people,' I say.

'Sorry, not in the real world,' says David. I guess he didn't read the correction.

To my disappointment we move on without calling the Bureau of Meteorology. Instead, I remind Joanne and David that many independent records have been used to construct the temperature graphs that they claim are 'corrupt'. It's not just one data set. I really doubt that four separate institutions made up of the world's top scientists would let measurements from weather stations close to urban heat sources destroy the integrity of the whole data set. Furthermore, scientists don't just look at *surface* atmospheric measurement stations to come to the view that the earth has warmed. They also use data from the oceans and the atmosphere—all of which show a long-term warming trend since observational data has been available.

'I mean, how stupid do you think climate scientists are . . . they *do* account for these things,' I say, frustrated.

'Well they have a failed theory on their hands,' says David. 'And they're concealing.'

'They're concealing?' I ask.

'Obviously,' says David.

It's an extreme claim, but David and Joanne have an explanation ready.

'A lot of money was starting to move into the field [of climate science],' says David, 'banking on the fact that we would get a carbon trading scheme of sorts. There were a lot of people, bureaucrats, people with careers that depended on it. I mean, you can't blame them for trying to find all the reasons why their story, their institution, their thing was right.'

'People have put in $79 billion, in fact the number's probably much higher than that now, to paying scientists to find a crisis,' adds Jo.

'So we reckon the climate scientists have a failed model on their hands and they're cheating and concealing the failure of their model. One of the ways they conceal the failure is to mis-measure the temperature,' David says.

The surface temperature records that David feels are 'mis-measured' are replicated in Figure 5.2. The graph shows surface temperature records from the three major scientific agencies that conduct independent measurements: NASA's Goddard Institute for Space Studies (GISS), the National Oceanic and Atmospheric

Figure 5.2 Surface temperature records from three major scientific agencies

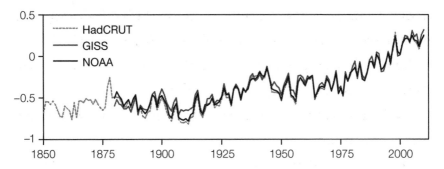

Source: Adapted from the Australian Academy of Sciences, *The Science of Climate Change: Questions and Answers*, report, August 2010

Administration's National Climatic Data Centre in the USA, and the Climate Research Institute at the UK's Hadley Centre (HadCRUT).

'So you're saying this line from NASA is wrong?' I ask as I point to the part of the graph that shows NASA's temperature record data.

'This line from NASA comes from this thermometer, in part,' says David, pointing at one of his photos of a 'corrupt' thermometer. 'It's wrong. And we don't trust them.' Later he adds, 'This [line] is out there on its own using those little corrupt thermometers and it's too warm. They're cheating.' It seems he didn't believe me when I said that temperature records are adjusted to take account of urban heat.

If you really believe that NASA and its data can't be trusted then I guess you don't use a GPS, I think. And you might want to ask some hard questions about the moon landing. But ultimately this debate comes down to more than who to trust. Joanne and David provide no evidence to negate the key facts that show human-induced climate change is happening. They haven't shown that humans don't emit greenhouse gases. They haven't shown that greenhouse gases don't trap heat. They haven't shown that the earth has not warmed since the Industrial Revolution. And they haven't shown that feedback mechanisms that worsen the initial warming don't exist. They won't put forward a theory of their own to account for climate change, and seem comfortable ignoring the only explanation that fits the observed, empirical facts.

'Non-believers don't have to prove anything,' wrote Joanne in her *Skeptics Handbook*. 'Skeptics are not asking the world for money or power.' I don't remember asking the world for money or power either. All I'm asking for is a chance to safeguard our climate, the basis for life on earth, from severe disruption. By contrast, professional sceptics like Joanne and David seem willing to risk the future of my generation without accepting they need to make a solid case for delaying cuts to carbon pollution.

I remind Joanne that the Intergovernmental Panel on Climate Change has said there's more than a 90 per cent probability that the climate change we're seeing today is mostly caused by humans. I don't even get *into* talking about the fact this number isn't 100 per cent because of the IPCC's extreme caution. Each IPCC member country—including blockers like Saudi Arabia, who try their very best to water things down—must sign off on reports. Even despite this in-built conservatism, the IPCC's recent report on extreme weather events revised the probability estimate for 'increases in the frequency and magnitude of warm daily temperature extremes, and decreases in cold extremes'. The new probability of these temperature extremes happening due to climate change is 99–100 per cent.

'If engineers told me that [a] car that I'm about to get into had a 90 per cent chance of crashing, I would not get in it,' I say.

'I'd ask the engineers for some evidence,' says Joanne.

It's a good line, great for television, but I do have to wonder: would she *really*? If Joanne is not a hundred per cent convinced, will she keep driving with an unsafe engine? If she's on a plane and the captain says there'll be a delay due to an engineering problem, would she march up to the cockpit and demand the plane take off, unless she's seen the report from the ground engineers herself?

In their book *Merchants of Doubt*, Naomi Oreskes and Erik Conway write that all social relations are trust relations. We trust other people to do things that we can't or don't want to do for ourselves. They use the example of a title search that is done when we purchase a house, pointing out that we trust the title search company to do it. We simply don't have the expertise to do it ourselves:

> If we don't trust others or don't want to relinquish control, we can often do things for ourselves. We can cook our own food, clean our own homes, do our own taxes, wash our own cars, even school our own children. But we can not do our own science. So it comes to this: we must trust our scientific experts on matters of science, because there isn't a workable alternative. And because scientists are not (in most cases) licenced, we need to pay attention to who the experts actually are—by asking questions about their credentials, their past and current research, the venues in which they are subjecting their claims to scrutiny, and the sources of financial support they are receiving.

The question that remains to be asked of Joanne and David is what happens if they're wrong? When facing the level of risk posed by climate change, how much knowledge do we need before we have enough to act? I have a friend who sums up the dilemma well: we could bet the house on the chance the sceptics are right. Or we could insure the house on the chance they're wrong.

On the off chance—or, according to the IPCC, the ninety per cent chance—that David and Joanne are wrong, most people agree it's sensible to take a conservative approach to the risk of climate change. The rational path is to take precautions to ensure the planet's climate stays within a range that provides relatively stable conditions for human civilisation.

'OK, let's use a really simple analogy,' I say. 'Most people take out car insurance every year. It's about $1000. You don't get to the end of the year and go, "Damn, I didn't crash my car."'

Joanne responds: 'So if the guys who are selling you car insurance are saying you need to spend $50,000 to insure your $50,000 car, then you'd be saying, hang on, why? And then they say, well, look at the accident statistics, and we say well let's look at them and they say, well we're withholding all the data, we don't give you the publicly available data on it.'

Joanne's response about the $50,000 car costing $50,000 to insure doesn't make much sense as a climate change analogy. First, just as no car costs the same to insure as what it's worth, it doesn't cost the entire value of the earth to save the earth. Second, it's much cheaper to take action to reduce carbon pollution now than to wait until the damage is done and try to adapt. The longer we wait, the more expensive it gets. The eminent economist Sir Nicholas Stern found that the economic impact of unmitigated climate change would be greater than both world wars and the Great Depression combined. Yet he estimated that successfully tackling global climate change would cost only one per cent of global GDP. He later revised this estimate to two per cent, because the world has taken so long to act.

But Joanne and David strongly disagree with this approach to risk management. They also disagree that they have anything to prove. In law, and usually in life, the burden of proof lies with those arguing against received wisdom. Joanne and David are clearly arguing against the mainstream scientific consensus. Virtually all climate scientists publishing peer-reviewed research in the area accept that climate change is real and caused by humans. The reality of human-induced climate change is also accepted wisdom among the world's national academies of science and royal scientific societies, and such organisations as the CSIRO, NASA and the Bureau of Meteorology. Even oil-dependent Saudi Arabia has had to grudgingly add its name to the list of nations endorsing the findings of the Intergovernmental Panel on Climate Change. But Joanne and David still refute that the onus is on them to prove anything.

There's a break in our discussion while David goes to pick up their three children. When we re-convene, I decide to discuss with Joanne the role her blog plays in generating hate towards climate scientists.

As one of the main online gathering places for the small but vocal number of Australian sceptics, the comments section of Jo's blog contains a lot of vitriol. One of the highlights as I checked her site is the following: 'Govt whore, The revolution IS coming. People have had a Gutfull of this global warming FRAUD and LIES! You can only push people so far before they bite back . . . HARD!!' This rise in anger among a certain group of climate sceptics has only escalated as the science has become more settled. Journalist Jo Chandler writes 'the cyber tirades are vicious, sometimes physically threatening, usually anonymous and very possibly orchestrated'.

Faced with the prospect of a price on carbon pollution targeting the top polluting industries in Australia, the rage has now spilled over from the internet into the real world. Just weeks before our meeting with Joanne and David, one of the world's most respected climate scientists was threatened by a man holding a noose a few metres in front of him at a conference in Melbourne. The scientist, Hans Schellnhuber, was visiting from Germany. After the incident, Schellnhuber told *The Australian* he believed it was only a matter of time before a climate scientist is murdered. 'As I tell my colleagues from time to time, "some day some madman will draw a pistol and shoot you". It will happen—to me or somebody else. I'm pretty sure about that,' he said.

While I've never been threatened with a noose, I have experienced some of the abuse and threats so commonly received by supporters of responsible climate change action. In 2009, when I was at the United Nations climate negotiations in Copenhagen, I opened my email one morning to find the following threat. I've quoted it in full, so you can appreciate both the context and the sender's enthusiasm for exclamation marks.

Dear Anna and Wendy

Haven't you listened to the radio (and the majority or DEMOCRATIC Australian's) like 2GB: Alan, Ray, Chris, Jason, and most of all Barry Wilshire, who has been rebutting your nonsense for over 20 years.

The game is up. You unintelligent, bureaucratic and moronic individuals!! If you think that causing Economic warfare and terrorism on regular Australians is the right thing to do . . . Prepare for an outright F*CKING REVOLUTION AND WAR!! !!!!!!!!!!!!!!! The Australian people - the same people who vote . . . . Don't want a fucking bar of your global conspiracy nonsense !!!!!!!!!! !!!

F*ck off!!!

Or you will be chased down the street with burning stakes and hung from your fucking neck, until you are dead, dead, dead!

Global warming, has now, according to Al Gore himself, if you are even intelligent enough to read . . . : has now turned to spiritualism, over global warming (oops, Al Gore 'now' says Climate Change . . . which is it AL???) Is it Global Warming, Climate Change, or have you gone spiritual on us now?

F*CK YOU LITTLE PIECES OF SH*T, SHOW YOURSELVES IN PUBLIC!!!

You won't be going home.

Bring it on!

Joanne has, in the past, insinuated that some of the death threats aren't actually made by sceptics, but instead sent by someone wanting to garner publicity for climate science. 'It's in quite a few people's interests to help those scientists win the sympathy of the crowd, and to distract the crowd with something non-scientific,' wrote Joanne on her blog. 'When the rock star fame is waning, a highly publicised death threat is a way to win sympathy and keep the celebrity factor rolling. It also makes your opponents look like criminals. Convenient, eh?'

I find this offensive. For years, climate scientists have been threatened, vilified and undermined by individuals and groups who refuse to accept the peer-reviewed consensus on climate change. When I show David and Joanne the death threat message that I received, David initially can't see the problem. 'Where's the death threat?' he asks. 'Whereabouts? Just point, whereabouts?' Then I read out 'you won't be coming home' and David realises he can't dismiss it. 'I think you're right to be upset about this,' he says.

'Yeah, well that kind of stuff is appalling and I would never condone statements that suggest anything violent,' says Joanne.

I'm glad we've found some common ground. But how does she reconcile this position with the level of hatred towards climate scientists on her blog? 'I worry that some of the debate that's happening on your blog may be encouraging things like this, in the comments section,' I say.

'If you see a comment that you dislike, please let me know,' says Joanne. 'I'd like to remove comments which suggest any form of violent intent and I try to keep on top of that,' she continues. 'As much as I can, I remove any threats of violence from both sides, because to me it's just wrong.'

'Sounds like we both agree,' I say.

By the end of the meeting I'm emotionally drained. Despite this, I'm glad I went ahead with it. I feel I'm beginning to understand where sceptics like Joanne and David are coming from. They see themselves as Galileo versus the Pope, an analogy they use several times. In their minds, they're representing the lone voices of sanity in a world that's gone mad.

But to me it's Joanne and David's arguments that are madness, not mine. They minimise the best-case scenarios of potential climate change impacts and claim everything will be fine. Then they exaggerate the worst-case scenario of the cost of dealing with the problem. They say that global warming is happening but carbon pollution isn't causing it, despite it being a basic principle of physics that CO_2 and other greenhouse gases trap heat. Then they do an about-face and argue global warming might not be happening anyway, because the temperature record is 'corrupt'. On that shaky basis, they're asking you and me to risk the future of our planet.

If Joanne and David were taking the same approach to any other subject, with the same lack of credentials, they'd be sidelined as quirky and eccentric outsiders. But, as hard as it feels, I know it's probably good for me to spend time with people like them. It's a reality check, helping me get inside the heads of those Australians who flat-out reject climate science. Joanne and David are determined and I shouldn't underestimate their reach. They're doing everything they can to, in David's words, 'undermine the credibility of the establishment climate scientists'.

As we leave Perth my mum calls me to find out how my day went. 'Not great,' I tell her. How do I put into words, in a way that she and others can understand, the fact that this climate fight is a race against time? All Joanne, David and Nick have to do to delay action on climate change is keep spreading the seeds of doubt. If they do this for long enough, the climate will pass tipping points and it will be too late. Nick will just keep repeating his favourite line: 'I remain to be convinced.' No question of risk, or responsibility, or the implications of refusing to be convinced. And if this works, and the world doesn't make substantial cuts to carbon pollution, in time, there's no going back.

Chapter Six

MEASUREMENTS IN THE SKY

IT'S A COUPLE OF HOURS BEFORE I'M DUE AT SYDNEY AIRPORT AND I'm frantically plucking jackets and thermals off the shelves at a camping store in Bondi Junction. As I drape woollen jumpers over my arm, the shop assistant doesn't look happy. I've arrived ten minutes before closing time and I look like I'm equipping myself for a major hiking expedition.

When I'd learned that Nick and I were going to Hawaii for the first overseas leg of the trip, I'd imagined myself on the beach. But this morning I discovered we'll be spending our time in subzero temperatures on top of the world's largest volcano. This means some repacking is in order: out with the summer dresses and in with the beanies. The assistant looks at her watch impatiently. Trust me, I say to myself, I'm doing this as quickly as possible. As much as I'd love to miss my long-haul flight with Nick, I know I can't. I buy as much as I can before the cash register closes. I don't imagine there'll be much in the way of duck-down jackets once we land in Hawaii.

A few hours later I'm at the airport. I find Max—and he is annoyed. Against strict instructions, Nick is not waiting in the check-in area where Max and Pete are allowed to film. Instead—and understandably—he's gone straight through the terminal to the Chairman's Lounge.

Even without Nick, we have sixteen bags between me, Max, Leo, Kate and Pete. Our group is the last to finish at check in: there's a lot of paperwork involved when you're travelling with expensive camera, sound and lighting gear. We stick out like a sore thumb among the excited families, loved-up honeymooners and sandy-haired surfers lining up for Hawaiian Airlines Flight 452 to Honolulu. There's a sense of grim determination to just get through the next four weeks. Like an extended trip to the dentist, I know it won't be fun but acknowledge it's important to do. I'm resolved to work as hard as I can to convey climate science to Nick and the television audience.

I'm jammed in the middle seat on the flight, with a noisy family on one side and Max on the other. I take a sleeping pill that isn't quite strong enough to make me sleep. At times I start to doze, but wake up with a start every few minutes. Eventually we arrive in Hawaii and, after transferring to a smaller plane, land on the Big Island. It's an impressive sight out the window. Dense, emerald green jungle covers one side of the island. Volcanic deserts dominate the other. Their irresistibly named peaks—Kilauea, Mauna Kea, Mauna Loa—tower above the cloud line like upside down ice-cream cones. I see waterfalls surrounded by great cliffs and black sand beaches fringed by long stretches of reef through clear blue-green water. It looks like the set of the TV show *Lost*, which I later learn is indeed filmed just one island over.

The next morning I throw open my curtains to see a beautiful sunrise, which soon turns into warm sunshine. Hello, Hawaii! I'm standing in one of the few US states on track to meet its goal of powering itself through forty per cent renewable energy and reducing energy demand by thirty per cent by 2030. According to *Forbes* magazine, Hawaii is becoming 'something of a test bed for renewable energy technologies', including geothermal energy, algae-based biofuels and smart-grid experiments. I venture downstairs to check out the water before the day's filming begins. Standing outside in the sun, I watch noisy holidaymakers dive into the ocean. After a few minutes I have to return to my room to get dressed, promising myself that one day I'll try to come back to Hawaii as a tourist. The concern I feel about the damage unchecked climate change could do to this beautiful archipelago helps me focus on the days of debate ahead as I layer a beige mountain-climbing duck-down jacket over a polar fleece, state-of-the-art thermals and thick socks.

Soon we're in a hired four-wheel drive on the road to Mauna Loa, the biggest volcano in the world. We're going there because of its scientific research station, set up over five decades ago by the late Professor Charles David Keeling. Look up the index of any climate change book under 'K' and you'll likely see several pages mentioning Keeling and his work. A distinguished scientist, Keeling was awarded the US National Medal of Science in 2002. This is the nation's highest award for lifetime achievement in scientific research. And although he passed away in 2005, climate scientists today still stand on Keeling's shoulders.

Nick and I both know that scientists have understood the nature of greenhouse gases since the 1800s. The English scientist John Tyndall discovered that greenhouse gases trap heat during basic physics experiments with water vapour and carbon CO_2 in the 1860s. This is accepted even by the wackiest of climate sceptics. Tyndall pondered whether these gases might have changed the earth's climate in the past, and whether they could do so in the future in a seminal lecture to the British Royal Society in 1863 entitled 'On Radiation Through the Earth's Atmosphere'. Greenhouse gases hold heat in our atmosphere, said Tyndall, and without them 'the warmth of our fields and gardens would pour itself unrequited into space, and the sun would rise upon an island held fast in the grip of frost'. So important was the role of greenhouse gases, he found, that slight changes in their atmospheric composition could bring about variations in climate. Little could he have imagined what lay ahead for future scientists, and the world.

After Tyndall came the Nobel prize–winning chemist Svante Arrhenius. In the late 1800s Arrhenius became the first person to investigate the effect that a doubling of CO_2 in the atmosphere would have on global climate. In his paper 'On the Influence of Carbonic Acid in the Air Upon the Temperature of the Ground', he put forward the view that as humans burned fossil fuels and released CO_2, we would warm the planet's atmosphere dramatically—up to 8 or 9 degrees Celsius in the Arctic regions.

The next big scientific leap from theory to real-world observation came in the late 1940s and 1950s. American scientists, flush with Cold War funding and given the freedom to work on a wide range of scientific problems, started to study the climate and look into emissions of greenhouse gases. It was one of these scientists, Charles Keeling, who gave us the measurements that made the world sit up and start to take notice.

Around the time Nick was born in the 1950s, Keeling had a hunch that humans were pumping enough greenhouse gases into the atmosphere to start heating the world, with consequences for rainfall, food production, sea-level rise and the natural systems on which humans rely. Scientists knew this could happen *in theory* from the work of Tyndall and Arrhenius. But the concentrations of these gases in the atmosphere weren't being measured, so no one knew if there was a real-world problem. Back then, many scientists thought that the ocean would soak up extra carbon pollution, thereby preventing it affecting the atmosphere.

Keeling started working with the Scripps Institution of Oceanography in San Diego. He decided, since no one was monitoring greenhouse gas concentrations in the atmosphere, it was up to him to do it. He considered the most remote places in the world—far away from city and traffic air pollution that could contaminate the results. Keeling selected two sites: Mauna Loa in Hawaii, and a South Pole station in Antarctica. He started measuring the 'atmospheric constituents that can change the Earth's climate' (as the brochure from the Mauna Loa research station neatly explains it) in 1958.

The question Keeling was trying to answer was simple: is human activity changing the chemical composition of the atmosphere? He quickly discovered that his measurements were fitting into a clear trend. The Scripps Institution website gives an insight into the impact of these early measurements. 'Within a few years of measurements,' it states, 'the Mauna Loa record had changed the notion of the atmospheric CO_2 increase from a matter of theory to a matter of fact.' One of Keeling's colleagues, Professor Roger Revelle, helped analyse the data and warned that human-induced climate change had begun. Revelle called the situation a 'large scale geophysical experiment'. One of the people who sat up and took notice was one of Revelle's students, Al Gore.

We now have five decades of data from the Mauna Loa Observatory and hundreds of other measuring stations around the world. Taken together, they paint a distressing picture of rapidly rising CO_2.

Carbon dioxide is now at its highest point in over one million years. The Keeling Curve, shown in Figure 6.1, has become an iconic symbol of the destructive impact of humans on the planet.

•

Today, Nick and I are on our way to see where Keeling's scientific breakthrough began. We travel in separate cars on our drive up the mountain. Max wants to separate us in order to limit our off-camera conversations. The scenery is too stunning for us to do much talking, anyway. As the run-down houses of downtown Hilo give way to scattered dwellings among thick tropical forests, we drive past enormous trees. Their wide trunks tower over caravans and colourful fibro shacks. We pass some waste land that a group of kids have taken over and turned into an impromptu soccer field. As we climb higher, the vegetation gets smaller and hardier. Stunted shrubs wither before they reach one metre tall, struggling to grow

Figure 6.1 The Keeling Curve

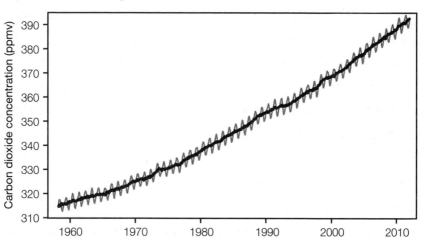

Source: Adapted from NOAA, www.esrl.noaa.gov/gmd/obop/mlo

in the black volcanic rock. We're entering a remarkable ecosystem, devoid of vegetation, known as a volcanic desert.

On the road to the research station, there are no other vehicles in sight. We arrive at the entrance, driving past a faded sign that reads 'visitor parking'. We're on an island in the middle of the world's biggest ocean, on top of the world's largest volcano, a three hour drive from the nearest town. I don't imagine they get many visitors. Stepping out of the car, I'm blasted by icy air. I've never been above the cloud line before, other than in an aircraft.

Two scientists called John and Aiden emerge from the building to greet us. John has been working at the observatory for eighteen years. Aiden, his assistant, is a Hawaiian local. After a few moments chatting, a young man named Ryan comes out to join us. He's a visiting grad student completing his PhD on laser radars—or, as they're called by the scientists, 'lydars'.

The research station consists of two small white buildings and a lot of unrecognisable equipment and instruments. I've never been to Antarctica, but this is how I've always imagined it would feel. It's freezing cold, the air is thin, and we're surrounded by a thick white blanket of clouds on every side of the mountain. We're 4000 metres above sea level.

John and Ryan usher us inside the larger of the two buildings. I'm grateful for its warmth. However, I know we'll be outside again soon, so I quickly add extra socks and thermal layers to my outfit and top it off with gloves. No one else is as cold as me and Nick jokes that I should love climate change since it's warming the planet. I smile. My poor circulation and perpetually cold fingers means it's a joke I've made many times myself.

Ryan is around my age and from Boulder, Colorado. He has happy, crinkly brown eyes and shoulder-length dark brown hair pulled back into a ponytail. He shows me the small kitchen. We share wedding planning stories: he got married just a month ago. Ryan's wife must be used to spending long periods without him as he lives for two months of the year in Greenland, and is now only stopping at Mauna Loa briefly on his way to a remote part of American Samoa. He is completing his PhD with the National Oceanic and Atmospheric Administration (NOAA), the highly respected scientific research organisation that's also an arm of the US government. As a mountain climber who loves the cold, being part of the measurement teams at remote research observatories scattered around the world seems to suit Ryan perfectly.

Ryan and I make tea while I explain what we're doing here. Nick is off exploring the rest of the building. 'This is a documentary where we go around the world and I try to change Nick's mind on the science of climate change, and he tries to change mine,' I say. 'I take Nick to meet scientists and experts who I think he'll listen to. He takes me to meet people who don't accept the science to try to change my mind, too.'

Ryan's disbelief is so extreme I wish Max or Pete were in the room to capture it on film. His eyes boggle and he almost starts laughing from the shock. 'Seriously? He doesn't accept climate science?' he asks. 'No, you're joking, right?' I explain that unfortunately this is not a joke. I tell Ryan that I hope being here at Mauna Loa might help make the scale of what's happening to our planet more real to Nick.

I explain that Nick was a former science minister in the Australian government, and Ryan's face changes from astonishment to bewilderment. Most of Ryan's life is spent in the scientific community in far-flung corners of the earth doing hands-on research. The fact that a former science minister with access to all the evidence in the world would refute the science of climate change is incomprehensible to him. Nick would probably say Ryan's reaction is evidence of scientific group-think but to me it's refreshing.

I've barely finished my cup of tea when Max makes us leave the warm kitchen and come back outside. First he films Aiden collecting samples of air. This involves some impressive-looking tubes and containers, but it's a simple process. The air goes into a flask, is flushed through it several times, and then captured when the flask is closed. The flasks are then sent to NOAA's headquarters in Colorado to be analysed and the information collected is added to data from other research observatories all over the world. NOAA receives over 190 flasks with air samples each week to analyse—around 10,000 per year.

I feel immensely privileged to have the opportunity to see these three thorough scientists collect the data that takes the measure of the planet. Max and Pete film me, Nick and John talking in front of the equipment about the work of the research station. The extreme wind conditions are causing a lot of sound trouble for Leo, but he perseveres. Our conversation in this magical place is eventually captured.

We know from the ice core data that when Nick was born in 1953, the concentration of CO_2 in the atmosphere was around 308 parts per million (ppm). In 1958, when Charles Keeling began his Mauna Loa measurements, CO_2 was up to 315 ppm. By the time I was born—exactly thirty years and one day after Nick—the number had risen to 340 ppm. And today, as Nick and I watch the air sample being collected, CO_2 is at 390 ppm. I find it strangely moving to be here witnessing these measurements. Here I am at the source of the data that first showed the world something disturbing was going on with the climate.

John is the chief scientist here. He has a beard, a moustache and a thick head of snow-white hair—although he usually covers it with a hat, given the high risk of sunburn at this altitude. He speaks calmly, slowly and deliberately. He seems so removed from the enormous controversy his work causes far away from here in the hallways of power.

John doesn't have any slick media lines. He hasn't prepared any sound bites. He's not comfortable in front of the camera. He'd rather keep his hat on, even if it casts shadows on his face. He looks at the camera with mild curiosity, not the suspicion that comes from being burned by the media spotlight in this ugly debate over climate science. I ask if he's involved in the politics of climate change in any way. He says no, he just sticks to the measurements. 'We take very great care in making them accurate,' he says. 'Accurate over the long term.'

John tells us that CO_2 levels are at the highest they've been in the past million years. The Keeling Curve has just kept going up and up and up since before the Industrial Revolution (280 ppm) to today (390 ppm). This is a thirty-eight per cent increase in CO_2, and has led to most of the 0.8 degrees of warming the world has already experienced.

John says there is 'no argument about the increase being [caused by] fossil fuels'. There are two reasons that this is uncontested. Firstly, Keeling—the forward-thinking scientist that he was—also measured the isotopic composition of CO_2. Isotopes are simply different atoms with the same chemical behaviour as each other, but with different masses. For us non-scientists, this means Keeling could analyse not just how *much* CO_2 was in the atmosphere, but also what *kind* of CO_2. By looking at it closely, scientists can tell the *source* of a molecule of CO_2: whether it came from burning fossil fuels, as opposed to, say, a volcano. Second, we can simply look at the historical records of human activities. We know that since the Industrial Revolution we've been using fossil fuels and burning forested land, both processes which convert organic carbon into CO_2. 'We're sure that the increase is due to fossil fuel burning,' says John. 'That is the only thing that works, that makes sense.'

Most climate sceptics don't dispute the Mauna Loa measurements showing there's more carbon pollution in the atmosphere. A few do, but they're on the fringe and are soon put in their place by more media-savvy sceptics. 'Every time the subject of CO_2 measurements comes up people raise all kinds of objections to the Mauna Loa measurements,' writes an annoyed sceptic on Anthony Watt's blog *Watts Up With That*. 'I'm about as skeptical as anyone I know. But I think that the Mauna Loa CO_2 measurements are arguably the best dataset in the field of climate science. I wouldn't waste time fighting to disprove them.'

Nick Minchin has obviously heeded that advice. He accepts the Keeling Curve as an accurate record of rising CO_2 and accepts that the levels of carbon pollution in the atmosphere are now at the highest they've been for the past million years. The debate for him is just about whether or not this is causing the planet to warm.

'When we have CO_2 levels at the highest point in the last million years . . .' I say.

'Might not be a problem,' replies Nick.

'And temperatures that are correlated to it, then . . .'

82

'Might not be a problem,' repeats Nick.

'Most scientists are saying it is,' I say.

'Well it's because of the models,' says John.

John is talking about the models that show how CO_2 traps heat in our climate system and how the positive feedback mechanisms then kick in to amplify the original warming. Figure 6.2 shows measurements of CO_2 levels, temperatures and projected CO_2 levels for the future if we don't cut carbon pollution. The earth is in the hottest decade since global temperature measurement records began in 1880. We have also now reached the highest concentration of CO_2 in the atmosphere for a million years. I don't understand how Nick can look at these two facts and say it 'might not be a problem'.

One thing that stands out in the Keeling Curve graph is the fact that the line recording increased CO_2 isn't straight—it's jagged. Why are there are all those little spikes going up and down by corresponding amounts each year? On the first day Nick and I spent together he'd said, 'Only three per cent of the emissions are human, ninety-seven per cent are actually natural.' This isn't a lie, but it's certainly misleading since it only counts one half of the carbon cycle. It simply has to do with the way plants suck up and then release CO_2 as the seasons change.

To understand the carbon cycle, imagine a bathtub being filled from a tap at an even rate, with a sinkhole where the water drains away at the same even rate. The amount of water pouring in from the tap drains out at the same rate.

The natural carbon cycle is very similar to this. Plants breathe in CO_2 during summer as they grow, and then exhale it as they shed their leaves in winter and the leaves decompose. Like the water in

Figure 6.2 Past and future CO$_2$ concentrations compared to past temperature

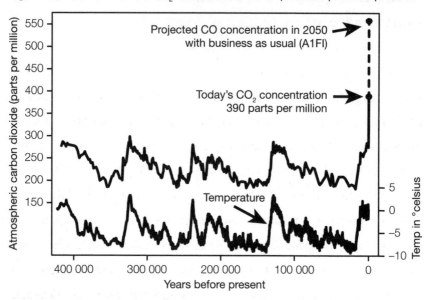

Source: Based on data from the Vostock ice core (Antarctica), the Law Dome ice core (East Antarctica) and Mauna Loa

the bathtub, carbon levels go up and down again. This happens every year like clockwork.

The carbon pollution humans are adding to this natural cycle is around three per cent extra every year. In our bathtub analogy, in addition to the water that goes in and out, you have three per cent more water going in each time the tap turns on. But the sinkhole remains the same size. Soon the bathtub overflows from all the extra water that keeps accumulating at a rate of 3 per cent more each year.

Standing in the corridor looking at the Keeling Curve on the wall, I'm reminded of something an Indigenous Australian man said to me six years ago about climate change. 'The planet is sick,' he said. 'She has a temperature.' To his eyes, the Keeling Curve looks like a hospital graph. To Nick, this would probably sound like one of his frequently used words, 'alarmism'. To me, the notion of a sick planet is spot on.

•

The day is getting late and we witness a spectacular sunset that bathes us all in a powerful, fiery golden light. Max films us talking until there's no longer enough light to keep going, then we move inside.

Our last bit of filming takes place when John turns on the lydar. This machine can only do its magic—measuring particulate pollution—at night. Particulates (sometimes called fine particles, or soot) are tiny bits of matter suspended in a gas or liquid. The name for the whole package—the particulates in their gas or liquid—is aerosols. I'm not talking about spray deodorant. Some particulates are produced naturally by volcanoes, dust storms and fires. Other particulates are generated by burning fossil fuels in power plants, cars and industrial processes. What you and I call 'air pollution', climate scientists call 'aerosols'.

John explains that the interesting thing about aerosols is that they mask warming. The *Science Daily* (June 2009) explains it succinctly: 'Diesel exhaust, industrial emissions, and the smoke from burning wood and brush eject bits of black carbon, usually in the form of soot, into the sky and form so-called "brown clouds" of smog.' These particles deflect sunlight back into space, creating a shading effect that cools the earth's surface. What this means is that the warming trend is probably less than it would have been without any aerosol pollution. However, this is only a temporary respite. When countries like China and India clean up their air quality, less sunlight will be reflected by brown smoggy clouds of soot. Then, the real temperature increases without aerosol masking will start to be measured and felt.

•

Our last conversation before we drive back to the hotel in Hilo involves another heated discussion about feedback mechanisms. Nick acknowledges the direct warming from carbon pollution alone, and focuses instead on challenging the existence of positive feedback mechanisms that amplify warming from CO_2. I'm glad we can agree that there is direct warming from CO_2 in the order of 1.2 degrees as a global average temperature rise. This is a point some other sceptics are yet to concede, which means they're refuting the existence of the greenhouse effect itself. At least Nick is rational enough to realise he'll never win the debate that way.

However, I can't see sense in Nick's position on feedback mechanisms. He argues that positive feedback mechanisms do not exist, and therefore don't worsen the initial temperature rise from CO_2. Going even further, he repeats the theory of Richard Lindzen, a climate sceptic scientist he's taking me to meet. Linzden's theory is that the earth has natural defence mechanisms—in the form of clouds—that stop it from warming.

'Lindzen is very much in the minority,' I say.

'Science isn't decided by consensus,' says Nick. 'It's not decided by minorities and majorities, nor is it decided by consensus.'

'Nor is it decided by you, Nick.'

'No, no,' he says. Then John, perhaps recognising that our conversation is going around in circles, changes the subject.

But the thing is, Nick isn't correct when he claims that science isn't done by consensus. It *is* done through consensus, which occurs through a process called peer review. In *Merchants of Doubt*, Naomi Oreskes and Erik Conway write about the fact that many climate sceptics have stepped outside the rules and guidelines of science that served to test the truth of scientific claims for over four centuries. Sceptics do this by going to the halls of public opinion rather than to the halls of science to make their claims:

> From its earliest days, science has been associated with institutions—the Accademia de Lincei, founded in 1609, the Royal Society in Britain, founded in 1666—because scholars understood that to create new knowledge they needed a means to test each other's claims . . . These were the origins of the institutional structures that we now take for granted in contemporary science: journals, conferences, and peer review, so that claims could be reported clearly and subject to rigorous scrutiny. Science has grown exponentially since the 1600s, but the basic idea has remained the same: scientific ideas must be supported by evidence, and subject to acceptance or rejected. Whatever the body of evidence is, both the idea and the evidence used to support it must be judged by a jury of one's scientific peers. Until a claim passes that judgment—a peer review—it is only that, just a claim.
>
> If the claim is rejected, the honest scientist is expected to accept that judgment, and move on to other things. In science, you don't get to keep harping on about a subject until your opponents give up in exhaustion. The he said/she said framework of modern journalism ignores this reality. We think that if someone disagrees, we should give that someone due consideration. We think it's only fair. What we don't understand is that in many cases, that person has already received his due consideration in the halls of science.

•

Nick, Kate and I drive back to Hilo though the volcanic desert under an enormous black sky. At one point we pull off the road to gaze in wonder at the stars. They're incredibly bright so far away from any human activity. The three of us stand there tilting our heads upwards. It is beautiful and still and the stars hang like vivid diamonds. It's moments like this that remind me that Nick and I have more in common than we do differences. We are both human beings, a tiny part of one species among millions, stumbling through life on a planet of which we are both in awe, in a universe too big for either of us to fully comprehend. I close my eyes and try to imprint the memory into the back of my brain.

On the drive home, Nick speaks about his career in politics. He is generous with his stories and anecdotes. He describes what it was like to be a cabinet minister; to divide his time between Canberra and Adelaide; and undertake an election campaign on the road.

We talk about the former leadership battle between John Howard and Peter Costello. Kate asks why John Howard had not handed the leadership over. Nick says that Howard honestly thought he would do a better job than Costello. I ask why he'd thought that. Nick says that he supposes it was convenient for Howard to think that. I smile—but from a place of sadness, not joy. It is convenient for people to think a lot of things, but it doesn't always mean they are right.

Chapter Seven

OUR FIRST CLIMATE SCIENTIST

TUESDAY MORNING DAWNS IN HAWAII. THE SUN IS SHINING WITH SUCH strength I can feel it before I even open the curtains. But there won't be any swimming for me this morning: I have to get ready for today's filming.

Under a patch of palm trees out the front of the hotel, Nick and I do a filmed debrief discussing yesterday's Mauna Loa trip. I'm wearing a green dress and the microphone battery pack strapped to my waist underneath makes me look like I have a small hunchback.

Despite the early hour, it's already searing hot. I look longingly at the ocean as a group of local kids dive off a pylon. Nick and I talk about what we learned yesterday while Pete and Max film us. Yesterday John had given us each a little glass bottle of Mauna Loa air, labelled with a picture of the Keeling Curve. I bring mine along to this morning's debrief and I wonder if seeing the accuracy of the measurements made an impact on Nick.

By now we've learned that CO_2 is at a point higher than it's been in the last million years. We also know that CO_2 traps heat, and that the temperature records show the world *is* getting warmer. Suddenly the science doesn't seem so complicated. Of course, there are some things that still need more research, such as the 'masking effect' of particulate pollution (the work Ryan is doing) and the exact magnitude of feedback mechanisms. But the basics are well established. Some people say science is never a hundred per cent certain. But after seeing early evidence linking smoking and lung cancer, a prominent British scientist wrote: 'All scientific work is liable to be upset or modified by advancing knowledge. That does not confer on us a freedom to ignore the knowledge we already have.' The fact that we don't know absolutely everything about the climate isn't an excuse to ignore what we do know with certainty.

Those facts that scientists *have* studied in great detail will be explained today by the head of the University of New South Wales' Climate Change Research Centre, Professor Matthew England. Matthew normally lives in Sydney but by lucky coincidence he's on a four-month research sabbatical based at the University of Hawaii.

It was tough for me to choose which scientist to ask to explain climate change fundamentals to Nick. Many people had suggested the director of NASA's Institute for Space Studies, James Hansen. Hansen was the scientist who first testified before the US Congress about climate change in the 1980s. He's often referred to as 'the grandfather of global warming', and not just because of his lifetime's work on the topic. It was the birth of Hansen's first grandchild that spurred him to take a more public role advocating action on climate change. Hansen has a lot to say about feedback mechanisms, which are the key sticking point for Nick. He's also passionate about the responsibility that the current generation of adults has to tackle climate change for the sake of their children. In a 2009 op-ed for *The Observer* he wrote:

Several times in Earth's long history rapid global warming of several degrees occurred, apparently spurred by amplifying feedbacks. In each case more than half of plant and animal species went extinct. New species came into being over tens and hundreds of thousands of years. But these are time scales and generations that we cannot imagine. If we drive our fellow species to extinction we will leave a far more desolate planet for our descendants than the world that we inherited from our elders . . . Young people are beginning to understand the situation. They want to know: will you join their side? Remember that history, and your children, will judge you.

Despite being in awe of Hansen's capacity to explain both the science and its implications, I decided not to ask him to be our 'climate science 101' guy for the documentary. I suspected Nick wouldn't be receptive to his arguments because Hansen had been publicly critical of Australia's role in the global coal industry. He'd even written to Australia's prime minister about the issue. It turns out I was right: when I bring up Hansen later, Nick says, 'He's a complete fraud and I wouldn't go and see him either.'

Once I'd ruled out Hansen, I'd started thinking about the best Australian scientist to explain the basic facts of climate change. Professor Matthew England fitted the bill perfectly: he's one of Australia's best climate scientists and a respected contributor to the Intergovernmental Panel on Climate Change (IPCC). Matthew's specialty is oceanography—a crucial area of climate research since about ninety per cent of the heat trapped by greenhouse gases ends up in the ocean.

Matthew meets us in the foyer of our hotel wearing a sky blue shirt that matches his bright eyes. His shaved head is as tanned as the rest of him and I remember he's a surfer and long-distance ocean swimmer. He tells me that the University of Hawaii is a great place for a sabbatical and that his three children are enjoying their new school. It's nice to finally meet Matthew in person—we'd only communicated online up to this point as he'd patiently answered all the questions I'd asked him about climate science.

We haven't organised a location to film our conversation yet, so Matthew and I take a taxi to the Hilo campus of the University of Hawaii. It's a small, friendly site scattered with tropical flowers and palm trees. After inspecting a few fluorescent-lit classrooms, a woman from the student centre offers us use of the art gallery. It's perfect—there's plenty of space and natural light. Nick, Pete, Max and Kate join us, carrying cardboard trays filled with coffee from the campus coffee shop. The drip-filter coffee at the hotel has been making everyone depressed.

Matthew introduces himself to Nick. Matthew recalls chatting to Nick while boarding a flight from Canberra once. They didn't talk about climate science, but today is Matthew's chance to make up for that and he's taking the responsibility seriously. Using his visiting professor status, Matthew's been able to borrow a flip chart and some coloured markers from the administration block. He uses these to draw a diagram of the greenhouse effect that looks like the one depicted in Figure 7.1.

As Matthew draws each aspect of the climate system, he describes it with reverence. 'We've got a planet that's about twenty-five per cent covered by land [and] seventy-five per cent by ocean,' he says as he colours the land masses green and the oceans blue. 'That's a great thing actually, because the oceans have a tremendous capacity to absorb heat.'

Figure 7.1 The greenhouse effect

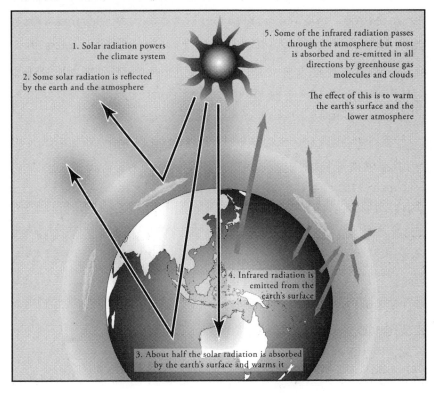

'Next up we've got to draw in the atmosphere, and that's just . . . a thin layer of gases,' he says as he concentrates on drawing a line around the earth. 'There are all sorts of gases there—oxygen and nitrogen, but of course also greenhouse gases—carbon dioxide, water vapour is also a greenhouse gas, methane and so on.'

Then Matthew moves on to drawing the sun, and the heat (or, in science-speak, 'radiation') that the sun emits. He explains that greenhouse gases trap heat. Since humans have been pumping out more greenhouse gases (as we saw yesterday at Mauna Loa) more heat has been trapped in the atmosphere instead of being radiated back into space. Matthew uses the analogy of a blanket:

Putting on an extra blanket—or two or three—progressively ramps up the heat. Your body is radiating heat outwards, the blanket is radiating the heat back down to you, so you get an enhancement of the heat trapping from the blanket. Greenhouse gases act in much the same way. They basically absorb and emit radiation, so they trap heat and they enhance the heat of the earth's surface . . . carbon dioxide is the big concern because it lives in the atmosphere for hundreds to thousands of years.

Nick agrees that Matthew's explanation of the way the greenhouse effect works is accurate. By this stage, we have established some common ground. Nick and I agree that human activities emit greenhouse gases. We agree that these gases trap heat and stop it escaping into space. We agree that, over time, temperatures have indeed increased. However, Nick prefers to look at short rather than long timelines. He likes to use the past decade to claim the earth has stopped warming, instead of looking at the long-term trend. I get frustrated when he does this. It's true that 2010 was only slightly hotter than 1998, and both vie with 2005 as the hottest years on record—but climate scientists say that to extrapolate trends in climate you must look at twenty- to thirty-year timeframes and more. Nick does know this and generally accepts that temperatures have increased since the Industrial Revolution.

We also both agree with the IPCC's assessment that (in the absence of any amplifying feedbacks) the average global temperature increase from a doubling of carbon pollution alone will be 1.2 degrees Celsius. We disagree on whether there are feedback mechanisms that will amplify this warming of 1.2 degrees. Climate scientists estimate that for every 1 degree of warming from carbon dioxide, there are around

3 degrees of warming from feedback mechanisms. As James Hansen says in his book *Storms of my Grandchildren*, 'Feedbacks are the guts of the climate problem . . . Feedbacks determine the magnitude of climate change.' But Nick doesn't accept this critical fact.

Nick and Matthew move on to a substantive discussion about amplifying feedbacks. Matthew explains that CO_2 has already increased thirty-eight per cent over pre-industrial levels, and this has led to a temperature increase of around 0.8 degrees Celsius. This is despite the significant masking effect of aerosols (particulate air pollution, which reflect heat back into space temporarily). This 0.8 degree Celsius temperature increase has also occurred despite the fact that the atmosphere requires several decades to fully heat up from each incremental rise in greenhouse gases. When climate scientists talk about feedback mechanisms they're trying to estimate how sensitive the climate system is to these feedbacks—hence the term 'climate sensitivity'.

Given we've had a 0.8 degrees increase with a thirty-eight per cent jump in CO_2 levels, what will the temperature increase be with a hundred per cent increase? It's worth noting that *when* this hundred per cent increase happens is entirely up to humanity. If our current fossil fuel–intensive emissions trajectory continues, we'll reach it as early as 2040.

Matthew tells Nick that with a hundred per cent increase in CO_2, most climate scientists expect the planet to warm by around 3 degrees Celsius. He explains that there's a range of uncertainty: 'It could be as high as 4.5, it could be as low as just under 2 degrees Celsius.'

'Depending on the assumptions you make about feedback,' says Nick.

'Depending on how the planet *decides* to feedback,' replies Matthew. 'We can't tell the planet how it's going to respond to greenhouse gases.' He explains how scientists arrived at their conclusion that climate sensitivity is around 3 degrees. 'We can look at past climates to get a very good idea as to how sensitive the system is to carbon dioxide,' he says. 'We can see these past records of the earth's orbit changing, and we see in response to that, greenhouse gases change, [then] temperature changes. So we can get at the sensitivity of global warming from the past climate records and from climate systems models. Those numbers *both* come at this value of 3 degrees Celsius as the best estimate, plus or minus about a degree and a half.'

Nick doesn't try to contradict Matthew straight away, even though disputing feedback mechanisms is the crux of his case to prove climate change isn't dangerous. Instead, Nick asks Matthew what the problem is with a 3 degree rise in global average temperature rise. Nick doesn't seem to think it's a big deal.

'Most people want to move to where it's warm,' says Nick. 'Everybody moves to Queensland and they move to Florida. And all the Russians and Canadians think it's fantastic it's going to warm up 3 degrees.'

'Although the Russians had a very unpleasant summer a couple of years back,' Matthew corrects him, referring to droughts and fires so severe that the Kremlin had to stop all wheat exports just to safeguard domestic food security. Matthew lists the climate change impacts we've seen so far at just a 0.8 degree warming: 'Extreme events, storms, hurricanes like Katrina and Yasi, drought cycles, bushfires . . . it's estimated that the cost of *this* level of warming is already significant,' he says. 'And so the big concern is if we go to 3 degrees Celsius—which is about four times that—then you start moving to extremes.' With extremes comes the possibility of tipping points. Matthew explains that when temperatures were raised in the past, sea level was tens of metres higher than today.

Now that 3 degrees doesn't sound so good after all, Nick goes back a step and challenges Matthew on whether feedback mechanisms *can* actually create a temperature rise of that scale.

'But . . . we mentioned Lindzen before,' says Nick, 'and he's the one who's most out there saying you're all *assuming* this positive feedback. It's not the CO_2 itself, it's its impact on water vapour, isn't that right?' he asks.

Water vapour is the gas phase of water and is continuously created by evaporation. This is a topic that sceptics sometimes bring up. Water vapour is a greenhouse gas, like carbon dioxide but with the ability to trap even more heat than CO_2. So, when sceptics talk about water vapour they imply that carbon pollution isn't a major problem. But this misses the point—the two gases are intricately related. As humans emit more CO_2, more heat is trapped in our atmosphere, leading to rising air, land and sea temperatures. This causes the ocean to heat up, sending—you guessed it—more water vapour into the sky. So the emission of one greenhouse gas, CO_2, actually leads to the release of water vapour (another greenhouse gas) from the ocean, which can trap even more heat, and the cycle continues. This is a 'positive feedback loop' but the results are less than positive—it significantly exacerbates global warming.

Matthew answers Nick's question about whether Richard Lindzen is right about positive feedback mechanisms not existing. Lindzen has put forward the view that water vapour will, instead of staying as vapour, change into the type of high clouds that reflect heat back into space and act as a kind of natural defence mechanism for climate change.

'The atmosphere gets more moist as you warm the planet. That's very true,' says Matthew. 'The atmosphere will get more moist, but water vapour is a greenhouse gas. So if the moisture forms nice big white fluffy clouds that reflect heat then that reduces the incoming heat, but plenty of studies have shown that the most *significant* impact of water vapour is to *increase* the greenhouse gas trapping . . . So for every degree Celsius warming you get from carbon dioxide, you can get up to another degree Celsius from water vapour feedbacks.'

Nick doesn't accept this. 'I thought that was still an area of substantial argument', he says.

'It's not nearly as debated as some people are trying to make out,' replies Matthew. Lindzen proposed his theory in the 1980s. 'When it first came on the scene everybody jumped at this idea and thought it makes a lot of physical sense . . . but of course then the research was done, people analysed the observations in more detail, they looked at more sophisticated models, and it's been since debunked.'

'In other words, you're completely confident there is this significant positive feedback which is the thing that's causing the dangerous warming?' asks Nick. He doesn't think that a 1.2 degrees Celsius rise in global average temperatures is a problem on its own. By that logic, it's only feedback mechanisms we have to worry about.

'No, positive feedback is part of the mix,' replies Matthew. 'Even without that you still get significant warming of the planet from greenhouse gases. You don't need the water vapour feedback to make this problem a problem. If you have the water vapour feedback, it can make this problem a *massive* problem.'

Matthew tries one last point to explain to Nick why scientists know that feedback mechanisms exist. He says that without feedback mechanisms, you don't have ice age cycles. In the past, he tells us, the size of the forcing that triggered ice ages was tiny compared to the response. Back then the initial warming was caused not by human emissions of CO_2 but probably by the earth's orbit around the sun changing course, or by changes in the intensity of the sun. 'So it seems from the past climate records that we can bump this system a little bit and get a big response,' says Matthew. 'At the moment we're bumping it by a big amount and it's going to be a big response.'

The point I am hoping Nick takes away from all this is that estimates of climate sensitivity are not just based on climate models. They're based on actual, legitimate data about how the climate system responded to temperature rises in the past. 'We know what that 100 part per million change can do to the global temperatures from the past records,' says Matthew. Any student of history knows that to ignore the lessons from the past can be a terrible mistake.

On that note, Matthew wraps up our meeting with a historical perspective. He talks about how former US president Jimmy Carter knew climate change was a problem in the 1970s. Matthew says:

He commissioned a report and the outcome was that the leading scientists of the day told him that the time to act on this issue was now. And it's amazing to me to see one of the world's most significant leaders of the day getting that advice. And of course, since that time scientists have written report after report and we've never found that advice to be wrong. We have said, hey, look, here's a better model, here's more observations, here's more data—but the basic physics

were known back in the early 1970s, enough to tell the world's most
significant leader of the day to have a good look at it.

Despite the frustration that Matthew must feel as one of those climate scientists writing 'report after report', he ends on an optimistic note. He believes we have a chance of avoiding dangerous climate change if we're able to harness the full potential of human creativity and ingenuity into solutions. 'It really is down to—there's no other way to put it—a warlike effort to mobilise a thriving economy away from fossil fuels and into low carbon technologies,' he says. 'It actually means a technological revolution. And that can lead to economic prosperity, but it needs such a significant and rapid scale of change.'

After we finish filming, Nick and I move to the waterfront to discuss what we've just heard. Overall, Nick was impressed with Matthew ('the most magnificent blue eyes I've ever seen') and commented on the way he didn't 'preach' about the science. It's true; Matthew is about the least patronising climate scientist you could imagine. He didn't display any anger towards Nick for the role he's played in obstructing action on climate change. He just explained the science clearly and answered Nick's questions thoroughly.

I ask Nick, 'Given the information that Matthew showed us today, is it responsible to say no, we shouldn't listen to this, no we shouldn't act?'

'Well, I don't like notions of responsible and not responsible,' Nick replies.

'I think the whole debate comes down to responsibility,' I say. We get into a debate about intergenerational equity. How is it responsible or fair to pass on the impacts of climate change to future generations? In terms of the timeframe, those of us alive today are the last generation with the ability to reduce emissions to levels that give us a chance to avoid climate tipping points. For us to wait and wait and wait until Nick is finally sure there is enough evidence to do something to reduce carbon pollution seems like the height of irresponsibility.

Of course, Nick sees it differently. We walk all the way along the length of the foreshore, Pete and Max in front of us with their cameras, as we argue the point. By the time we get back to the hotel we still haven't resolved anything. I head to my room and punch my pillow—just once, but hard—to vent some of my frustration.

After dinner, I get a chance to debrief with Matthew. We sit in oversized couches in the brightly lit hotel foyer under fans that try in vain to bat away the heat. I sip a glass of water and stifle a few yawns. It's been a long two days. Matthew and I both agree that Nick listened to what Matthew had to say. He clearly warmed to Matthew as a person, but still isn't convinced that mainstream climate science is accurate.

Matthew and I talk about the nature of the climate change threat. Because carbon pollution stays in the atmosphere for centuries and the climate system is a slow-moving beast, the impact of carbon pollution emitted thirty years ago is only now being felt today. This means that the impact of the carbon pollution we emit today will be felt by our children and grandchildren.

'We need a human psychologist here,' suggests Matthew, 'because we are very good at responding to swift, immediate threats. And climate change . . . a psychologist would say it's not happening quickly enough for humans to process it as a genuine threat.' What he's saying is that, like any other animal, humans respond to clear and present danger. In a 2006 article, Harvard psychologist and bestselling author Dan Gilbert argued:

> No-one seems to care about the upcoming attack on the World Trade Center site. Why? Because it won't involve villains with box cutters. Instead, it will involve melting ice sheets that swell the oceans and turn that particular block of lower Manhattan into an aquarium. The odds of this happening in the next few decades are better than the odds that a disgruntled Saudi will sneak onto an airplane and detonate a shoe bomb. And yet our government will spend billions of dollars this year to prevent global terrorism and . . . well, essentially nothing to prevent global warming. Why are we less worried about the more likely disaster? Because the human brain evolved to respond to threats that have four features—features that terrorism has and that global warming lacks.

Gilbert argues that first, climate change isn't an intentional evil designed to hurt us. 'If climate change had been visited on us by a brutal dictator or an evil empire,' he writes, 'the war on warming would be this nation's top priority.' Second, climate change doesn't violate our moral sensibilities. 'Although all human societies have moral rules about food and sex, none has a moral rule about atmospheric chemistry,' he says. 'The fact is that if climate change were caused by gay sex, or by the practice of eating kittens, millions of protesters would be massing in the streets.'

Third, says Gilbert, we see climate change as a threat to our futures rather than our afternoons. 'Like all animals, people are quick to respond to clear and *present* danger, which is why it takes us just a few milliseconds to duck when a wayward baseball comes speeding toward our eyes.' Climate change is a much longer-term threat.

And lastly, he makes the same point as Matthew England. The rate of change is slow enough (in human terms) to go undetected. 'If the low hum of a refrigerator were to increase in pitch over the course of several weeks, the appliance could be singing soprano by the end of the month and no one would be the wiser [until it breaks down].' Since the climate system changes slowly and gradually—until we reach tipping points, at least—many people don't notice there is a problem until it's too late.

As someone trying to convince others to take notice of climate change, I am constantly puzzled to see people respond efficiently to problems that meet the criteria Gilbert outlines, yet ignore the biggest threat facing our country's environment and economy. Australia has proven in the past few years that we can deal compassionately with floods, fires, droughts and cyclones. Yet in their aftermath, we stop one step short as a nation from thinking about how to deal with climate change. It's a terrible irony, given the extent to which scientists predict climate change will increase and worsen extreme weather events.

The human brain is an incredible organ, but we need to evolve one or two steps further when it comes to our ability to process danger. The greatest threats can be hard to physically see but have implications for decades down the line. If only humanity had chalked up more practice dealing with threats of time scales longer than a sports match before we were thrown the biggest curve ball of all.

Chapter Eight

SMALL GRAPH, BIG TROUBLE

Climate science is now so advanced that we can anticipate the kind of event that may, if we do not reduce the stream of greenhouse-gas pollution, initiate the end of the great 'us' that is our global civilisation. With no warning, a gargantuan ice sheet will begin to collapse. It will mark the beginning of an irreversible process and, even if the initial rise in sea level it causes is just a few centimetres, it will herald an abandonment of our coasts, for the ice must continue to melt and collapse, albeit erratically, until there is no more. It will be impossible to put a time scale on the flooding, but Shanghai, London, New York and most other coastal cities must suffer partial or total abandonment, over weeks or decades or centuries. With economies in ruins and infrastructure drowned, we will then all be on The Road.

Tim Flannery, *Here On Earth*

I'VE BEEN TALKING WITH NICK ABOUT ICE. NOT THE DRUG CAUSING trouble on the streets of Sydney, but rather the pure, white substance striking fear into the hearts of glaciologists and oceanographers around the globe. The NASA project that measures climate change impacts on Antarctica is called the Gravity Recovery and Climate Experiment (GRACE). This project is staffed by highly trained, deeply committed scientists who risk their lives working in unfathomable conditions at the two poles at the ends of the earth. Their job is to find answers in the stories unfolding in the Arctic and Antarctic, sifting through data from the ancient past to help us make decisions about the future.

Thinking back to my own past, one of my strongest memories is of sitting in a pew singing hymns up the back of St Joseph's Church. I went to a Catholic primary school and the weekly services were mandatory. I didn't mind. Sometimes we would sing my favourite hymn: 'Aha-ma-azing Grace, how swee-eet the sound,' I would belt out at the top of my lungs, too young to be self-conscious about singing. 'I once wa-as lost, but no-ow I'm found; was bli-ind but no-ow I seeeeee.' As an overly studious kid, the songs and the stories about revelation appealed to me. Somehow my young mind interpreted the idea of suddenly being struck by Grace/The Truth/ some deeper understanding of the world as a kind of shortcut to knowledge brought about by a flash of lightning. I wouldn't have to study any more! I would just *know*! Everything would change in the blink of an eye! To my great disappointment, as I grew older I didn't experience any moments of Divine Revelation. But I still believed in Grace. I saw it all around me in the beauty of the natural world and the kindness of strangers.

When scientists examine the bubbles trapped in ancient ice cores for information about our past climate, I romantically (and probably erroneously) imagine it as a kind of modern 'Amazing Grace' moment. This is why I love the choice of acronym for NASA's Antarctica project.

But this time the ice cores don't contain good news. They predict a mostly frightening future. Many climate scientists, including NASA's James Hansen, believe that the ice cores tell us more than climate models about how sensitive the climate is to warming from CO_2. It's the ice core data, in Hansen's view, that shows that climate sensitivity is around 3 degrees. This means that for every 1 degree of warming from CO_2, there will be a total warming of something like 3 degrees due to the feedback mechanisms.

Thinking about this, I cast a silent prayer. God, if you do exist, and care about Creation, now would be the time to step in. I haven't given up on miracles, but I believe we're going to have to save ourselves. I think back to a piece of graffiti I used to see every morning at the beach where I surfed before school: *There's no justice, there's just us.* We're the last generation with a chance to avoid those ice sheets melting and the devastating chain reaction that follows.

I'm thinking all of this watching the lights of San Francisco glitter through fog as we land late Wednesday night. As we drive over the iconic bridge, Max pokes his head, then his upper body, through the sunroof hole. He films the city skyline against the night sky as the wind flaps violently in his face and howls around his body into the car. Eventually we arrive, exhausted, at our hotel.

Max hasn't yet told Nick and me who we're here to meet, so the next day I get my own revelation: it's Professor Richard Muller from the University of California Berkeley. Neither Nick nor I selected Muller as one of our interviewees. Max and the production team chose him as a 'neutral' participant whom neither Nick nor I will be happy with, but who will add a lot to the discussion.

Muller is not a climate scientist but rather a physicist. Nevertheless, he has taken a keen interest in the climate debate and is characterised as a sceptic in the articles I find. Nobel prize–winning economist Paul Krugman wrote in the *New York Times* that Muller's 'climate-skeptic credentials are pretty strong: he has denounced both Al Gore and my

[Krugman's] colleague Tom Friedman as "exaggerators," and he has participated in a number of attacks on climate research, including the witch hunt over innocuous e-mails from British climate researchers.'

Having Muller leaning towards their side was a boon to the sceptics. He's a prolific researcher, recipient of the McArthur 'genius' fellowship, and part of a select group of scientists called JASON (named after a hero of Greek mythology), who advise the US government on matters of defence, science and technology (a JASON report in 1979 predicted human-induced global warming and urged the government of the day to take action).

Muller's latest project comes off the back of his being publicly critical of the temperature record data. His reason? Like Joanne Codling and David Evans, Muller was dubious about the locations of some of the temperature measurement stations. Muller has the same kind of photos Jo and David threw down on their kitchen table of weather stations near airports and air conditioners.

Unlike most sceptics, though, Muller decided that if you're going to criticise something you should base it on data—not just photos. To this end he assembled a crack team of scientists and data geeks. Their goal was to reconstruct the temperature record from scratch, trawling through and plotting measurements from weather stations the world over. They named the work the Berkeley Earth Surface Temperature Project (BEST).

The climate sceptic blogosphere got excited about Muller's project. Anthony Watts, the man who started the whole trend of taking photos of weather stations next to air conditioners, went to meet Muller. Watts was impressed. He shared his data with Muller's team and praised the project's methodology. On his blog Watts declared himself 'prepared to accept whatever result they produce, even if it proves my premise wrong'. Support also came in from Koch Industries, one of the top ten air polluters in the United States. Greenpeace calls the Koch Foundation 'a financial kingpin of climate science denial and clean energy opposition'. The Koch Foundation is the largest single donor towards Muller's study. It gave $150,000 towards the project's $620,000 budget.

Climate campaigners were less excited about Muller's research, given his previous attacks on climate scientists. Blogger Joe Romm pointed out that Muller's outside interests included serving as president and chief scientist at a consulting agency, Muller & Associates, which advises energy companies in areas such as enhanced oil recovery and underground coal gasification.

·

The night before meeting Muller, I don't sleep a wink. The evening begins well, with dinner at a funky Japanese restaurant with one of my bridesmaids, Grace, and her husband. They live in San Francisco and it's great to be able to see them again. For a few hours I feel like a normal young person again. After a cursory mention of the documentary, we're soon chatting about life, love and wedding planning. After dinner I reluctantly bid them farewell and head back to the hotel for a good night's sleep before tomorrow's meeting with Muller. At least, that's my plan.

As I lie there, the responsibilities of this project feel heavy. I think about how the documentary could help people understand the science better. But it could also confuse them further. I think about how it could change people's minds and clarify the basic science, or push them into Nick's web of denial. I get scared, suddenly, that this whole project is a mistake. Despite my fears, I feel better when

I think about how the debate over the science in Australia can't get any worse than it currently is. This turbulence in my head is not useful for anyone, and I need to get on with doing further research on Professor Muller.

I use the hotel's computer to print out article after article about Muller and his work, then go back to bed armed with a highlighter and notebook. I read through them all. I take notes. I highlight. I compile a list of things he's got wrong, and things he's got right. I finally finish, with a list of items to talk to him about tomorrow. I'd hoped that by now my mind would be relieved about finally getting a rest. But my thoughts return to the intricacies of the climate science that I've learned about so far, especially the secrets of the ice. Tipping points, ice cores, Antarctica, the Arctic, satellite data, glacial cycles. I've opened a tap in my brain that I just can't turn off.

This goes on all night. At about 5 am, an hour before I have to get up, I suddenly have a realisation. It's actually the closest thing to an epiphany I've ever had.

I've always had a huge sense of responsibility for people around me, and for myself. This has led to one of my biggest faults—I'm kind of a control freak. I try to make sure everything is as close to perfect as possible. Then I get stressed when things don't always turn out the way I plan.

Going into this journey with Nick, I probably over-prepared. Every evening after my paid job finished I would spend five or six hours researching which spokespeople I should choose, looking into the spokespeople I assumed Nick would choose, studying climate science and preparing talking points on each aspect of the issue. All in an effort to exert control over a process that deep down I knew I'd have very little control over. I mean, it's reality TV! How much of a say do I have in how it comes together in the editing suite? None.

My revelation from my state of sleep-deprived angst was this: I can't really control this journey at all. I can't control whether Nick changes his mind. I can't control what the spokespeople on the side of the science say, or how they come across on camera. I can't control how this journey is portrayed by the director, or even how *I* am portrayed. By the end of the show they'll have hundreds of hours of footage that need to be compressed into just one hour, which means they can make me out to look like anything they want me to look like. All I can control is how I act on a day-to-day basis.

All I have control over today (and it's a slim hold on it, I can assure you) is making sure I keep it together during our meeting with Muller. Don't say something stupid because you're delirious with exhaustion, I tell myself sternly.

•

Muller meets us in Sproul Plaza, outside Sproul Hall at the centre of the Berkeley campus. It's a symbolic place for Muller. He was arrested here in 1964 during a student protest supporting free speech and academic freedom.

Almost five decades later, Muller is still at Berkeley, now as a professor of physics. When we meet him, I can still imagine the young student troublemaker determined to cause a fuss no matter the consequences. He's likeable but more than a touch arrogant, with a broad and quirky smile. His lined face bristles with untamed whiskers, his clothes sport more than a few food stains, and he carries a small backpack. He talks animatedly, and a lot. I tell him I've read that he was the most popular lecturer on campus. He immediately corrects me: 'No, I was the *best* lecturer on campus.' You can tell he's used to being the biggest personality in the room.

As we walk through the plaza, Max gets Muller to talk about his arrest back in 1964, and then segue into the problems he has with climate change science. It's clear to me the angle Muller's going for: free-thinking genius still standing up for free speech and what's right, no matter the cost. He's trying to compare the free speech and academic freedom battles of the 1960s with the climate sceptics' claim that their speech is 'oppressed' because their point of view is rejected by the scientific community.

Muller talks about how Anthony Watts' concerns about the location of some temperature measurement stations and the impact on the overall temperature records resonated with him. 'There were issues of the urban heat islands,' he says. 'There is warming that's taking place in the world just from the fact that people are burning, creating heat, they have air conditioners, they have things like that. How much has that contributed to global warming? Now these are legitimate questions . . . I became somewhat of a sceptic at that time just because these were legitimate questions that really demanded answers.'

I should stress here that there are already three independent temperature records used by the scientific community: one from NASA, one from the National Oceanic and Atmospheric Administration (the organisation where Ryan, who we met in Hawaii, works) and one from the Climate Research Unit at the UK's Hadley Centre. Professor Muller decided they weren't to be trusted and that he would create a new temperature record from scratch. In doing so, he was essentially arguing that three separate existing records, all showing the same thing, were potentially wrong.

I'm keen for Muller to talk about the findings of his research, but Max keeps bringing him back to the problems Muller has with climate science. It's the ABC's dream come true. Years ago, I'd been told by a Triple J journalist: 'We can interview you if you're able to find us a young climate sceptic to also interview.' (This is ridiculous in many ways, not least of all that it's extremely hard to *find* young climate sceptics that aren't pushing a certain political agenda.) The ABC is constantly trying to present climate change as a 'balanced' 50:50 debate. Now they've found someone—Muller—to argue both 'sides' from within the same body!

Max keeps prompting Muller to share his criticisms of climate science, and I feel like Max is essentially doing Nick's job for him. But there's nothing I can do, especially since I'm all Zen after my realisation about control a few hours ago. I listen patiently but fume inside as Muller trashes his colleagues in the scientific community with allegations that have been debunked many times.

For example, one of Muller's favourite complaints is about a graph by scientist Michael Mann. It's termed 'the hockey stick graph' (because of its shape). Professor Mann used tree-ring data combined with information from ice cores, corals and other physical proxies to create a graph that reconstructs temperature records for the last thousand years. He merged these proxies with the records from temperature measurement instruments.

Muller had been extremely critical of the hockey stick graph after thousands of emails were stolen from the Climate Research Unit, including many about the graph. He also points to the work of Steve McIntyre, who 'discovered that this hockey stick is based on a computer mistake'. McIntyre is a mathematician, not a scientist.

He is also a former minerals prospector, a semi-retired mining consultant and chair of the board of a mining company.

The reality is that while the hockey stick graph is not perfect, the data that Professor Mann (now director of Penn State University's Earth System Science Center) used to create his graph has been found to be solid. The *Guardian* reported that:

> *Upwards of a dozen studies, using different statistical techniques or different combinations of proxy records, have produced reconstructions broadly similar to the original hockey stick. These reconstructions all have a hockey stick shaft and blade. While the shaft is not always as flat as Mann's version, it is present. Almost all support the main claim in the IPCC summary: that the 1990s was then probably the warmest decade for 1000 years.*

In 2006, findings were published from a US National Academy of Sciences inquiry into the hockey stick graph. It upheld most of Mann's findings, albeit with some caveats:

> *There is sufficient evidence . . . of past surface temperatures to say with a high level of confidence that the last few decades of the 20th century were warmer than any comparable period in the last 400 years. Less confidence can be placed in proxy-based reconstructions of surface temperatures for AD 900 to 1600, although the available proxy evidence does indicate that many locations were warmer during the past 25 years than during any other 25-year period since 900.*

So while the hockey stick graph was not faultless, it was broadly accurate. Given that more than a dozen subsequent scientific papers produced reconstructions similar to the original graph, the level of criticism lumped on Professor Mann, including from Muller, seems to have been unfair.

But we're not here to talk about the hockey stick graph. We're here to talk about the results from Muller's work on the Berkeley Earth Surface Temperature Project (BEST). Muller pulls several graphs out of a manila folder in his backpack. He says, 'In order to get to the bottom of these issues I felt we needed a good, solid, powerful scientific team consisting of people who had never taken a political stand on global warming but knew how to handle complex data.' This team of eight used records from all 39,000 weather stations around the world (operated by NASA, NOAA and the Hadley Centre), comprising 1.2 billion data points.

Muller explained how they went about it. 'We looked at the effects of the extra heating that takes place in cities which is not part of the global warming. We looked at the station quality bias . . . We looked at the data manipulation. We did not adjust the data ever so we were able to do this and put together a record.'

Muller's results were no surprise to climate scientists, but a huge shock to the sceptics that had endorsed his work. Muller tells Nick and me that 'we're getting basically the same result as other people . . . it's amazingly similar.' In fact, Muller found that in the past forty years the land has warmed nearly four times faster than it did in the last century. In the key scientific paper summarising the work, Muller wrote that the team found the global average land surface temperature has increased by 0.9 degrees Celsius since the 1950s, plus or minus 0.04 degrees Celsius. 'Our analysis suggests a degree of global land–surface warming during the anthropogenic era that is consistent with prior work (e.g. NOAA),' says Muller's paper.

Nick and I are lucky enough to be hearing Muller's preliminary data before it's been fully published. Muller will later go on to tell US media outlet MSNBC that that 'we're getting very steep warming' and 'we are dumping enough carbon dioxide into the atmosphere that we're working in a dangerous realm'.

Figure 8.1 is the graph showing Muller's findings. It plots the temperature records used by NOAA and the Hadley Climate Research Centre (CRUTEM), and compares them with Muller's Berkeley analysis.

Figure 8.1 Muller's findings compared with records used by NOAA and the Hadley Research Centre

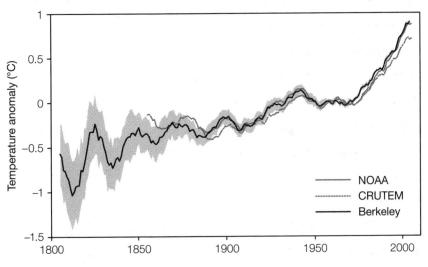

Source: Adapted from BEST, http://berkeleyearth.org/analysis

Nick doesn't really react to the big-picture findings Muller has just revealed. Instead, he asks questions about the details. 'Are these preliminary?' 'How far back does the data go?' 'Is this all a land-based measurement?' He doesn't argue with Muller. After all, Muller has already looked at the sceptic arguments over temperature records, found them flawed and answered them with this study. Nick seems to take Muller's experience—of doubting and double-checking the science and finding it accurate—in his stride.

Muller tells Nick that he shouldn't dismiss the data from NASA and the other scientific organisations. 'When it came out so close what it made me appreciate was that the other scientific teams were actually better at *doing* science than explaining to others what they had done,' he says. 'They were really good scientists who had worked hard and been very careful. They couldn't convince the sceptics that they were careful, but they *had* been careful.'

Muller believes that this finding should stop sceptics from arguing the temperature record is corrupt. I doubt this will be the case, but Muller seems confident. 'I believe we can convince the sceptics we were careful,' he says. 'Because of the way we handled this, with the openness that we're doing, with the details that we're doing, because we respect the sceptics . . . I believe we can convince any sceptic who is open-minded.'

I ask Nick, who claims to be an open-minded sceptic, whether seeing Professor Muller's research has changed his mind. 'Well I've never denied that we've gone through a warming phase,' Nick says. This doesn't seem to fit with the fact that Nick's first choice of spokespeople—Jo and David—made the temperature record a core tenet of their attack on climate science. Nick continues: 'Both the satellite and the land-based data show that subject to the sort of issues Richard's raised that there's been an increase in average global temperature from the '70s onwards.'

Nick goes on to focus on the past decade, which although it is clearly the warmest decade since records began, doesn't show much increase when you compare 1998 with 2010 (which are about the same). But as I've mentioned previously, climate data is measured and discussed in thirty-year trends. With Muller's graph in front of us, it's not hard to spot what the trend is: up, up and up.

Nick does try to argue one point with Muller, bringing up the fact that Muller's temperature measurements are for land-surface temperatures only. 'Seventy-five per cent of the planet's ocean and you're only measuring the twenty-five per cent that's land,' says Nick. Muller replies bluntly: 'We live on the land and it gets hotter on the land.' In reality, the ocean is warming too. The ocean holds most of the heat from global warming, but because it's so deep it is slower to heat up than earth's surface.

We wrap up our discussion by talking about the chances of the world acting with sufficient speed to solve climate change. 'It's not easy to do something about it. It doesn't matter what we do in the United States unless we affect the poor people of the world, somehow get them to emit less,' Muller says.

His argument is that even though right now countries like Australia and the United States are by far the largest emitters per person, China's and India's emissions are rapidly growing due to their large populations. China and India feel it's well within their rights to develop the same fossil-fuel intensive methods that Western countries used. For developing nations it's a case of 'the West caused it, so it's only fair the West act'.

This international relations hurdle can be overcome by action: actions speak louder than words, after all. If a grandfather who has been a smoker his whole life learns about the health risks and tries to convince his grandson not to take up smoking, he's much more likely to succeed if he gives up smoking too. If the grandfather keeps smoking while attempting to convince his grandson that smoking is a bad idea, he's more likely to fail. It's for this reason we'll never have a comprehensive international treaty to reduce global emissions unless industrialised countries like Australia (one of the highest emitters of carbon pollution per person) act on climate change with courage and conviction.

Muller seems at a loss to suggest how climate change can be solved. He's hoping for a new technology, and thinks that China and India will never choose to voluntarily limit their carbon pollution. But he's neither a development economist nor a technologist. Perhaps he's lost faith in the power of the human race to mobilise everything we have to solve crises when we finally realise we're under threat. Too many people move straight from denial to despair without having the hope and determination that we need to solve the problem.

After the conversation, Max and Pete go to film some shots of Muller riding his motorbike around campus. I wait to be filmed for a debrief, perched on the edge of a concrete flowerbed watching the happenings in Sproul Plaza. There's a dishevelled forty-something man with a few missing teeth terrorising hapless students with crazed environmental messages as they walk through the plaza. He's in great pain about what's happening to the environment. 'The planet is dying,' he yells at the top of his voice. 'You are the iPod generation! You care more about connection with your iPod than connection with humans or the planet!' The students carefully avoid him and avert their eyes. It's sad to see someone be so aggressive in the name of a good cause. He's certainly not giving Berkeley greenies a good name.

The shouting man prompts me to think about the challenges of trying to communicate the critical urgency of climate change to a public that often isn't interested. Well-meaning communications experts often urge campaigners like me to downplay the impacts of climate change. Their fear is that the public will get so scared they'll shut off, and people like Nick will succeed in labelling campaigners 'alarmists'. Many major environmental groups have heeded these warnings, focusing on the exciting potential of clean energy and green jobs instead of the scariness of climate change. The problem, however, is that these communications experts often have no understanding of how dire the situation actually is. So while I have no urge to join the man in front of me in shouting at people, I do question whether downplaying the risks is really the best strategy.

Max films me for a few minutes: he wants my impressions of Muller. I'm still exhausted and craving sleep, so I don't say much for fear of stumbling over my words. But I make my key point: Muller used to be a sceptic but now understands that climate change is happening and that temperatures are at the highest they've been in over 400 years. When humans started industrialising, temperatures started rising, and we've just finished the hottest decade since records began.

The data was enough to convince Muller to change his mind on climate science—but not Nick.

Chapter Nine

SAN DIEGO:
DOUBT IS OUR PRODUCT

THE NEXT MORNING WE FLY TO SAN DIEGO. IT'S ONE OF THE FEW places on our itinerary I've never been before, so I'm eager to explore. But the production company's travel agent has different ideas. They've booked us into a hotel just off the freeway, within walking distance of nothing but arterial roads. Kate and I go inside and manage to wrangle us out of the booking. We didn't come all the way to San Diego to stare out a window at American drivers commuting to satellite suburbs.

Our crew moves to a hotel in the city centre instead. It's a charming old building, formerly a bank. My room looks over the main street. A hot dry wind blows through the wide boulevard. I imagine an era long ago. San Diego used to be part of Mexico, and later flourished in the boom times of the Californian gold rush. By all accounts, the streets were filled with gold prospectors as they stopped here to purchase supplies and trade stories on their way to the Santa Clara mines.

The woman at reception tells me that the Mexican border is only a twenty-five minute drive away. It's so close I'm tempted to go to Tijuana for dinner. Max says he'll come with me. But instead, we end up filming set-up shots of Nick and me in a convertible driving along the San Diego coast. It's a truly bizarre moment: roof down, hair blowing in the wind, gorgeous car, the sun setting . . . and I'm sitting next to Nick Minchin talking about climate change. As always, the filming takes hours longer than expected. By the time we get back to the hotel there's definitely no time to go to the other side of town, let alone Mexico and back. Dinner is downgraded from a trip to Tijuana to a nondescript Mexican restaurant just off San Diego's main drag. Around us the streets fill rapidly with rowdy locals warming up for a night of partying.

After dinner, the filming must go on. Nick and I sit in the lobby of the hotel. We try to look natural perched on an oversized couch in front of the bright stage lights that Max and Pete have set up. We talk about the person we'll be meeting tomorrow, Professor Naomi Oreskes from the University of California, San Diego.

There's a quote by CS Lewis in Naomi's book, *Merchants of Doubt: How a Handful of Scientists Obscured the Truth on Issues from Tobacco Smoke to Global Warming,* that I want to show Nick. 'A belief in invisible cats cannot be logically disproven. It does, however, tell us a lot about the views of the person who holds it.' I read it to Nick, and explain it reminds me of the Australian climate sceptic Ian Plimer's claim that invisible, undetected underwater volcanoes are responsible for climate change (rather than human carbon pollution). Nick laughs, tells me he loves CS Lewis, and says he'll put it to Plimer. At least Nick has a sense of humour. It's not always as hard being around him as I'd expected it to be.

Naomi Oreskes is a scientist, a former mining geologist, and now a specialist in the history of science. Her book, co-authored with Erik Conway, also a historian of science, delves into the scientific consensus on climate change. Nick's first choice of spokesperson, Joanne Codling, argues that 'there is no consensus, there never was, and it wouldn't prove anything even if there had been'. Naomi's work has tackled the question of whether consensus around human-induced climate change exists in the scientific community. And, she went on to ask, if there is consensus, why is there such confusion among the public? Her book answers these questions, and when I read it a year ago I was transfixed by the story it told.

To understand the modern-day 'debate' about climate change we have to cast our minds back almost four decades, around the time the young Richard Muller was getting arrested in the Berkeley sit-in. In 1965 Charles Keeling and Roger Revelle (the two scientists I learned about at Mauna Loa) wrote in a Science Advisory Committee report to the White House that 'by the year 2000, there will be about 25% more carbon dioxide in our atmosphere than at present, and this will modify the heat balance of the atmosphere to such an extent that marked changes in climate could occur'. In 1979 the US National Academy of Sciences wrote that 'a plethora of studies from diverse sources indicates a consensus that climate changes will result from man's combustion of fossil fuels and changes in land use'.

Two years later, in 1981, the World Meteorological Association and the UN Environment Program established the Intergovernmental Panel on Climate Change (IPCC). This panel was established with the goal of summarising the state of the science. So in the early 1980s, business and government lost any excuse of 'not knowing' about climate change. But obviously not everyone was happy about the implication that humans would need to reduce our reliance on polluting forms of energy.

Naomi's work traces the origins of climate denialism. As I explained earlier, climate denialism is different to genuine scepticism in climate science and is instead based on a refusal to accept evidence. She traced its origin to three men who founded the George C Marshall Institute, a think tank located just outside Washington DC.

In the May 2011 edition of *Cosmos*, Naomi writes that 'for many years, scientists at the institute have either denied the reality of global warming; insisted that if there is warming, it was not caused by human activities; or, as they continue to today, insist that scientific uncertainties are too great to warrant government regulation'.

The Marshall Institute was founded by three Cold War physicists called Robert Jastrow, William Nierenberg and Frederick Seitz. These physicists built their careers working on the atomic bomb, the hydrogen bomb and other nuclear weapons delivery systems. They then worked together on the Reagan Administration's Strategic Defense Initiative (known as the Star Wars missile shield). The project was so controversial that 6500 American scientists and engineers signed a petition boycotting strategic defence funds. (Many feared it would re-ignite a nuclear arms race.)

In 1979, the story took a twist. Fred Seitz accepted a new role as a consultant for the RJ Reynolds Tobacco Corporation. The world now knows beyond reasonable doubt that smoking and passive smoking are harmful to our health. Court documents prove that the tobacco companies knew that too, but employed people like Seitz to cast doubt on the science. The 1969 tobacco memo that inspired the title of Naomi's book (*Merchants of Doubt*) read: 'Doubt is our product since it is the best means of competing with the "body of fact" that exists in the mind of the general public.'

By 1989 the Cold War had ended, and the three physicists (Jastrow, Nierenberg and Seitz) found a new enemy: environmentalism. Using the lessons from the tobacco industry, they followed the same path of casting doubt on the science. They didn't have to prove that climate change *wasn't* happening, just argue it wasn't yet settled.

Why did they do this? Naomi's research showed the motivation for Nierenberg, Seitz and Jastrow was a fear of environmental extremism, or exaggeration of environmental threats by people with a left-wing agenda. Naomi writes that, on this basis, they and other scientists with similar ideological views argued against 'the reality of acid rain, the harmful effects of the chemical DDT, the severity of the ozone hole and, of course, the human causes of global warming. In every one of these cases we see this small group of physicists denying the severity of these problems.'

In each case—the ozone hole, DDT, acid rain and climate change—Naomi found the same pattern. Scientists were working outside their field of expertise and misrepresenting the actual evidence to insist it was just not settled enough to justify the government doing anything about these problems. Naomi also discovered 'personal attacks on leading scientists, stealing private emails, and pressuring journalists to write "balanced" stories, giving equal weight to the industry position'.

Naomi noted the tactic of finding or cultivating 'a tiny handful of dissenting scientists—three physicists in America, two geologists in Australia, one climate scientist in New Zealand—and promoting them on television, radio and in print media to create the impression of real scientific debate'. We still see these tactics alive and working well in the climate fight today. A small minority of individuals have managed to trick large numbers of people into thinking no scientific consensus exists. A 2009 study by Professor of Earth Science at the University of Illinois, Chicago, Peter Doran, and his graduage student Maggie Kendall Zimmermann found that only fifty-eight per cent of the American public answered yes to the question 'do you think human activity is a significant contributing factor in changing global mean temperatures?' In contrast, 97.5 per cent of climatologists actively publishing in the area responded yes.

In the 1990s, the original three anti-climate science crusaders were joined by another contrarian, S. Fred Singer. Also a Cold War physicist, in the 1980s Singer had worked to cast doubt on the significance of acid rain before joining the others in arguing against the scientific consensus over the ozone hole and global warming. Singer also worked with Philip Morris to defend tobacco against the scientific claims about harm caused by passive smoking. Despite an expert panel concluding that passive smoking was responsible for more than 3000 additional adult cancer deaths and up to 300,000 additional cases of bronchitis and pneumonia in infants and young children each year in the United States, Singer vigorously argued that passive smoking should not be regulated. 'If we do not carefully delineate the government's role in regulating dangers, there is essentially no limit to how much government can ultimately control our lives,' he wrote in 1993.

As Naomi puts it, Singer seems to share a suspicion that 'environmentalists are really socialists in disguise'. For example, when writing about the ozone hole in 1989 he said, 'There are probably those with hidden agendas of their own, not just to save the environment, but to change our economic system. Some of these coercive utopians are socialists.'

This is the background research I try to summarise for Nick as we sit on the couch in the hotel foyer. Naomi's work shows that the 'debate' over the climate crisis was manufactured by the same people who brought the world 'safe' cigarettes'. I bring this up, and Nick and I talk about smoking. Nick clarifies and defends his position. He accepts that smoking is linked to health problems. 'I have always accepted that smoking will kill you and that's why I've never smoked,' he says. 'But I've always taken this libertarian view that if you want to kill yourself by smoking well, you know, we can all choose our own way to go.'

However, Nick rejects that there's a scientific consensus around the fact that *passive* smoking is harmful. He'd laid this view out clearly in 1995 when he was part of a Senate committee investigating the cost of tobacco-related illnesses. While the majority report of the committee favoured a raft of regulatory measures to reduce smoking, Nick had opposed many of the regulations, claiming the tobacco industry was already over-regulated.

In the report, Nick challenged the medical and scientific orthodoxy on the health impacts of passive smoking. 'Senator Minchin wishes to record his dissent from the committee's statements that it believes cigarettes are addictive and that passive smoking causes a number of adverse health effects for non-smokers,' the committee's minority report says. 'Senator Minchin believes these claims [the harmful effects of passive smoking] are not yet conclusively proved . . . there is insufficient evidence to link passive smoking with a range of adverse health effects.' *The Australian* reported in December 2009: 'To support his claims, Senator Minchin drew on a study commissioned by the Tobacco Institute of Australia that "concluded the data did not support a causal relationship between exposure to environmental tobacco smoke (ETS) and lung cancer or heart disease in adults".' It's telling, I think, that Nick drew on reports from the tobacco industry rather than doctors or medical associations.

Six years later, Nick still seems to hold this position. Nick claims that journalists have conflated his two positions (believing that smoking is unsafe but *passive* smoking is safe), in an attempt to smear his reputation. Personally, I think the fact that Nick holds the belief that it's safe to inhale second-hand smoke is damaging enough for his reputation. The evidence linking passive smoking to health problems is just as strong as the evidence for the damaging effects of first-hand smoke. The same carcinogens in first-hand smoke are present in second-hand smoke. Nick doesn't mind if people smoke—it doesn't challenge his world view. It's their individual right to kill themselves! However, accepting the evidence for passive smoking leads to *regulations* to protect people in public spaces like restaurants and bars. Nick doesn't support government intervention, so more regulation is not attractive to him. Perhaps he accepts science that doesn't challenge his world view but rejects science that does.

I bid Nick goodnight and go to sleep, listening to the sounds of the partygoers hitting the clubs and bars in the streets below. Most are probably oblivious to the warnings scientists have published about how climate change is likely to impact California, and with it, San Diego. The expectations for this area include more mega-fires, less rainfall, more drought and sea level rises that will do considerable damage to California's spectacular coastline. In the worst-case scenario, sea level rise could inundate coastal cities including parts of San Diego. This sounds alarming but it's not impossible. If the entire West Antarctic ice sheet melts, sea levels would rise six to seven metres. But even best-case scenario show major losses, which is why the former governor of California, Arnold Schwarzenegger, took such a strong stance on climate change. In 2005 he announced aggressive emission reduction targets for the state, saying: 'The debate is over. We know the science, we see the threat and we know the time for action is now.' California's emissions trading scheme starts in 2012 and its chief architect, Linda Adams, recently visited Canberra to share lessons on the scheme's design with the Australian government.

•

In the morning, we drive out on the vast, sprawling highway towards Torrey Pines State Natural Reserve. If you look at San Diego from the air or on Google Earth you'll see great swathes of highway cutting through the surrounding hills like a thick black spider web. The road to Mexico goes one way, the road to Los Angeles the other. Another road, leading east, goes through Arizona and will take you to Texas and beyond.

Naomi arrives in shorts and a green jacket, wearing hiking boots and pink socks. With her short dark hair and broad smile, she looks very much the like the mining geologist she once was—practical, strong and smart. She flew into San Diego last night, returning from a conference in Europe. The jetlag hasn't dampened her spirits, and she's enthusiastic and engaging.

I can tell that Nick has warmed to Naomi when he expresses concern about where she's standing. We're on the edge of a cliff overlooking the San Diego coastline and some incredible rock formations. It's slightly precarious and Nick is clearly worried that one of us will lose our footing and plunge to our death. It's an apt analogy for where humanity stands in relation to climate change. We're desperately trying to have a rational conversation while in imminent danger of falling over the edge of a cliff to tipping points and runaway climate change.

Naomi has a lot of experience with rocks, cliffs and other geological formations. As a young scientist she worked in South Australia (Nick's home state) for three years. She was a mining geologist with Western Mining Corporation working in uranium exploration. She'd even considered, at the request of a friend, running for the South Australian state government as a Liberal Party candidate. It would have been fascinating, I think, if Nick and Naomi had been in the same branch of the same political party.

Instead, Naomi moved back to the United States and forged a distinguished career in science and later in the history of science. Now Nick and I stand alongside her on the cliff and she gestures towards a white pier down the coast to our left. 'There's my office,' she says, pointing to the famous Scripps Institution of Oceanography within the University of California. Scientists at Scripps have measured the average surface temperature of the water at the end of the pier since 1950, and say it's increased by almost 3 degrees Celsius since then.

So far Nick's been reluctant to argue the climate science with my spokespeople. He usually prefers to mostly listen and ask questions. But Naomi is blunt, lively and down to earth, and she quickly draws Nick out of his shell.

Naomi's first question puts the burden of proof back onto Nick. 'What's lacking for you in the science that makes you feel like you're not convinced?' she asks. 'We have every major academy around the globe, Nobel prize–winning scientists, tens of thousands of scientific journals, published and peer-reviewed papers across the globe, scientists who are conservatives, liberals, Democrats, Republicans, Tories, an incredible diversity of people—so what's missing for you?'

I'm expecting Nick to trot out the one piece of scientific evidence that *would* convince him climate change was caused by humans and that feedback mechanisms are real. But he stumbles. 'I've been in politics a long time and I suppose my natural disposition is one of conservatism and scepticism . . . I try and examine all the evidence,' he says. And then he just goes back to the 'remain to be convinced' line. Naomi presses on, again asking what specifically is lacking in the science for Nick.

Nick tries to argue that he's not convinced because a lot of scientists aren't convinced. 'It remains . . . the case that there are a very significant number of major scientists—' he starts. Naomi corrects him: 'No, that's not true. It doesn't remain the case. There's a very, very tiny number of scientists, most of whom aren't actually climate scientists. This is what I've studied for the past five years.'

Nick tries a new tack. He argues the earth has experienced 'cyclical warming' ('I'm old enough to remember the global cooling scare,' he says) and a 'levelling off' of warming since 1998. 'That's not true,' says Naomi, 'but why don't you finish and then we'll respond to some of the things that you said.' Nick argues that despite the 'levelling off' of warming, CO_2 keeps rising. I think he's trying to suggest that CO_2 and warming are not related.

'One of the things that intrigues me is that we know CO_2 is a very minor greenhouse gas—' he begins to say. Naomi interrupts: 'I guess I have to stop you now because you're saying things that are just actually not true. It's not a minor greenhouse gas. Base CO_2 in the atmosphere before the Industrial Revolution was 280 parts per million. It's an incredibly small amount of stuff and yet that small amount of stuff is what makes our earth liveable. Without those 280 parts per million the earth would be completely uninhabitable. So what we know is that carbon dioxide is an incredibly powerful, sensitive greenhouse gas, that small amounts have radical consequences for climate. So if you say it's a minor greenhouse gas you're really flying in the face of what we know from basic physics.'

'Well my understanding is as a component of all the greenhouse gases it's something like three to five per cent, that water vapour is the overwhelmingly the greenhouse gas,' says Nick. 'It's water vapour that essentially is the primary controller.'

'Nick, just because there's something that's a small amount doesn't mean it can't have a big impact on something,' I say. Carbon dioxide lasts in the atmosphere for hundreds of years longer than water vapour. Water vapour does trap heat. However, water vapour is only emitted when the oceans heat up because of the initial warming from CO_2.

'I think we're going off on a track,' says Naomi. 'You said quite a few things that I would argue are not actually correct.'

'Sure,' says Nick. 'I'm here to listen to you rather than me trying to convince you.'

'That's alright,' says Naomi. 'I mean, you can try to convince me if you want.'

Nick claims that there is natural variability in the climate throughout the history of the earth, and that modern climate change might just be part of this natural variation. 'We've had a warming period since about 1680 out of the little Ice Age,' he says. 'The planet's been warming *without* the influence of human-induced CO_2 emissions. That's consistent with the history of the planet which is one of substantial climate change quite separate from human activity.'

'Correct,' says Naomi. 'We know there is natural variability. Nobody would deny that—but you can have more than one thing going on in the world at once. So what we have *now* is the natural variability—that's quite well understood scientifically—and then we've imposed upon that an *additional* cause. It's an additional factor that wasn't there before humans started burning fossil fuels and that's the driver of increased greenhouse gases in the atmosphere.' Naomi points out that humans have been burning fossil fuels since the Industrial Revolution, 'but in an accelerated way in the last fifty years or so. That is the period in which the natural cycle has been altered. If you look at the natural climate records, what you would expect is that actually the natural variability ought to be moving us into a cooling trend—but it's not. Instead what we've actually done is we've reversed the natural variability.'

Naomi tries to impress on us the scale of this change in the climate system. 'That's a huge transformation,' she says. 'This was the insight that Roger Revelle wrote about right down there at the end of that pier fifty years ago. He said humans have become a geological agent. We have reached the point in history where the activities of humans are now on the same scale as natural geophysical phenomenon.'

Naomi explains that Revelle's conclusion was controversial at the time. The prevailing logic in geology was that the earth is vast and changes slowly and humans were just another trivial species among many. Revelle realised that this was no longer true, based in part on his colleague Charles Keeling's measurements. And this was how the modern-day field of climate science began.

Nick seems to run out of things to say on the science. When the burden of proof is flipped and he has to try to justify why he rejects the science, he struggles. Perhaps this is why he fights my characterisation of the burden of proof so hard. We'd had an instructive conversation earlier in the day about this. I had been explaining that, given the fact that climate science is accepted by every major scientific academy in the world, it's up to Nick to prove to me that it's *not* caused by humans—as opposed to me proving to him that it *is*. 'You can't reverse the onus of proof,' Nick had said.

'We can,' I replied, 'because the consequences of not acting are so disastrous.' Also, it's the job of scientists to do science. So if a politician rejects their work I'd argue it's up to the politician to give a decent reason why. Here's some of our conversation, recorded on Max's tape:

NICK: *It's like a criminal trial where there is simply, what do they call it, not correlation, what's the . . .*

ME: *Reasonable doubt?*

NICK: *What's that term in criminal law for the evidence? No, no, no, for the evidence that is . . .*

ME: *Admissible? All the admissible evidence points to the fact that beyond reasonable doubt climate change is happening, it's caused by humans. So for you to convince the world not to act you would have to show beyond reasonable doubt that it's not happening.*

NICK: *That's applying the criminal test. No, no, no for those who want to overturn the current state of affairs and the current structure of our economy and our civilisation.*

ME: *The current state of affairs is a climate that's habitable for humans.*

NICK: *Yeah, but for those who say you can't keep doing what you're doing, you have to radically change what you're doing. The onus rests on them.*

ME: *But what you're doing is advocating a course of action that could radically change our whole society, our economy, our planet. That's much more radical than wanting to keep a stable climate.*

NICK: *No, no, no, but you're saying the status quo cannot continue, you're the one advocating radical change.*

ME: *No, actually I don't see it that way. The way that I see it is that I'm advocating the status quo climate. I want to keep clean air, clean water, a stable climate.*

NICK: *And you want human beings to radically change their behaviour.*

Nick's last point hints at something we're about to explore with Naomi. By this stage of our conversation, we've moved on from the science. We're now talking about Naomi's research on consensus in the scientific community. Naomi tells us about her study in 2004 that analysed the scientific literature on the question of global climate change.

'We had a hypothesis based on what was in the US National Academy of Science's review of climate science,' she says, 'in which they had said that the American scientific community agreed with the Intergovernmental Panel on Climate Change, that most of the observed warming over the past fifty years is likely to have been due to the increase in greenhouse gas concentrations. Our question was how many papers published in peer-reviewed scientific literature *disagreed* with that?' Naomi wanted to find out how much dissent actually existed.

'We did an analysis using the Institute for Scientific Information database,' says Naomi. 'We did a keyword search on global climate change.' Her results were a huge blow to the people arguing the science was not settled. 'We found that *not one* [article] presented evidence—either theoretical or observational—to refute that National Academy conclusion,' she says. 'That was really a shock to me because . . . the people who don't agree get a lot of attention in the media.'

Naomi describes seeing these results as a turning point for her. 'The only debate really, in the scientific community, was how fast was it happening, how quickly would changes occur in the future and how extensive were the impacts already,' she says. This led her to ask a simple but powerful question. 'When I realised that the science was so secure, I started wondering why do we all have this impression that there's a raging debate? Why did the media present it as a debate?'

It was this question that prompted Naomi down a research path that led, surprisingly, to the tobacco industry. 'In the tobacco legacy documents we found extensive evidence of how the tobacco industry had developed an organised campaign,' says Naomi. 'What we learned from our research was that that same strategy has been applied to climate change. The media have been bombarded with press releases, with white papers, with reports that assert that there are questions about the science, that assert that we don't really know, that there isn't a consensus, that we don't know about the cloud feedbacks and that we're not really sure how significant CO_2 is.' Naomi pauses for a moment and looks at Nick. 'All of these allegations—almost none of which are supported by peer-reviewed science—sound plausible or sound like an issue that a reasonable person ought to be open minded about.'

The question Naomi grappled with was why Jastrow, Nierenberg and Seitz would participate in this kind of campaign. 'Why would they deny the scientific work, collected by their own colleagues, work they clearly understood because they were pretty smart guys? Why would they reject it?' asks Naomi. After all, they were *scientists,* not tobacco industry executives.

Naomi explains that these men, the 'merchants of doubt' to which her book title refers, were Cold War warriors. They 'saw the defence of the "free world"—and therefore free markets—as an extension of their life's work'. Unlike the tobacco executives, their primary motivation was not personal financial gain. They were driven by something much bigger: defending free market fundamentalism and campaigning against government regulation. 'We looked for what they themselves said about it,' says Naomi, 'and it was all about this issue of what this problem meant for government control of our lives.'

But what does rejecting climate science have to do with defending free markets, I wonder? Well, to accept that climate change is a problem involves accepting that we need to do something to solve it. What Jastrow, Nierenberg and Seitz feared is that this 'something' would involve over-zealous government regulation of fossil fuels and, by implication, the economy and people's lives.

'In a way these guys made what philosophers call a category mistake,' says Naomi.

'They had a legitimate concern but then they moved it into an area where it didn't really belong . . . then they began working with the tobacco industry, the fossil fuel industry and building this whole network of denial in the United States.' Since modern industrial society depends on fossil fuels, to argue that fossil fuels were hurting the planet was very confronting. 'Lots of people were willing to [come] on board because at the end of the day, to accept that climate change is real is upsetting,' says Naomi. 'It's frightening, it's troublesome and it has implications that are rather distressing.'

The reason climate denial has been largely propagated in America by pro–free market, anti-regulation think tanks is the threat of corporate regulation as part of a package to solve climate change, explains Naomi. These organisations include the Competitive Enterprise Institute, the Cato Institute, the Heartland Institute and the Institute for Public Affairs in Australia. These groups are funded, explains Naomi in her *Cosmos* article, 'by regulated industries that produce products that have serious negative consequences not addressed by the free markets'. These industries include tobacco, fossil fuels, mining and the chemical industry.

Earlier in the conversation, Naomi had asked Nick why he rejected the science. Nick had answered by talking about what the implications of the science being correct would mean. 'I remain to be convinced that we have a major problem *which requires massive industrial change*.' Nick's earlier comment earlier to me ('you want human beings to radically change their behaviour') echoes in my mind, giving me a greater understanding of his fears. It seems Nick approaches climate change in a similar way to the original group of physicists who founded the Marshall Institute. Naomi has a name for it: implicatory denial. 'We don't like the implications of what the climate science tells us *not about the earth*, but what it tells us *about us*, about the way we live,' she says.

Implicatory denial. I think Naomi has hit the nail on the head. It's not *really* the science that Nick has a problem with. It's the implications of the science for our economy, our society and, in Nick's case, what it says about the need for government regulation in the market.

In 2006 the British government commissioned the highly respected economist Sir Nicholas Stern to write a report about the economic impacts of climate change. He returned with findings that must have struck fear into the heart of every neoliberal free-marketeer around the world:

> *Climate change is a result of the greatest market failure the world has seen. The evidence on the seriousness of the risks from inaction or delayed action is now overwhelming. We risk damages on a scale larger than the two world wars of the last century. The problem is global and the response must be a collaboration on a global scale.*

For someone like Nick, this is devastating. Market failure? Global scale collaboration? The whole economic conservative worldview is under attack. What could be greater anathema to those whose ideological foundations rest on the mantra of small government? Suddenly, we're looking at countries cooperating to reduce emissions, and Sir Nicholas Stern is calling for rich countries to lead the way. Is Stern arguing for some kind of international communist takeover led by the United Nations?

Stern's prescription for how to avoid the risk of enormous economic damage from climate change include: 'adopting ambitious emissions reduction targets; encouraging effective market mechanisms; supporting programmes to combat deforestation; promoting rapid technological progress to mitigate the effects of climate change; and honouring . . . aid commitments to the developing world.'

These kinds of ideas and initiatives are horrifying to people like Nick, just as the thought of them was horrifying to the three physicists who had spent their lives fighting the Cold War.

Nick doesn't seem convinced that the reason he rejects climate science is to do with his political and economic ideology. This is despite the fact he fits the ideological profile of Jastrow, Nierenberg,

Seitz and Singer. 'I just don't like the state telling me what I can and can't do,' he'd said to me last night as we discussed laws around smoking. Today he argues that science and economics are linked. 'You can't separate them,' he says. 'And so for people like me having been in . . . government, if very significant economic change is being proposed as a mechanism for dealing with a problem you've got to be damn sure we've got a real problem here, and a problem we can actually do something about.'

Naomi acknowledges Nick's fears about economic change. 'The steps that we will need to take to address this problem are very, very dramatic. We're talking about a complete transformation of the energy system, which is the basis of our economic activity,' she says. 'So I think you're a hundred per cent correct to say we want to make sure that we move forward with confidence that we're basing our decisions on good information. But here's the thing: you're not doing that.'

Naomi challenges Nick about the source of his views. 'You're accepting information from people who have their own reasons for denying the science, motivations that do *not* come out of problems in the scientific evidence,' she says. 'And it's actually boxing you into a corner of rejecting the science and delaying the sensible discussion about what to do about it. And the problem is getting worse.'

Naomi makes her final point, appealing to Nick's own values. 'The longer we wait, the more likely it is that we're going to have to do things that you don't like,' she tells him. 'The people I've studied, one of the things they feared was a massive government intrusion in our private lives—and I agree with them! I don't want the government telling me what to do. But the longer we wait and the worse this problem gets, the more likely that becomes,' she concludes. 'So that's for me the aspect of your situation that actually defeats your own goals.'

We wrap up the meeting because we're soon due to the airport. Naomi has given us a lot to think about on the long plane trip to Boston.

'So have you changed your mind?' I ask Nick cheekily as we're leaving.

'Good summary,' Nick says of the way Naomi recounted the history of the campaign against climate science. 'We're on a journey here.'

'Give him time,' says Naomi. She turns to Nick and smiles. 'She's young and impatient. You have a few more weeks, you know'.

On the drive to San Diego airport I check some Australian news websites on my phone. The headlines feature mounting opposition to the government's planned price on carbon pollution. They remind me that there is more than one type of denial. Literal denial—arguing that human-induced climate change actually isn't happening—is the obvious kind.

But then there's second type of denial. A large number of people understand that climate change is happening, and that it's bad—but are in denial about the fact that our economy must change in order to solve it. A case in point for this is the large number of Australians who say they care about climate change compared with the large number opposing the price on carbon pollution.

This is what happens when we know about climate change but don't integrate that knowledge into everyday life and political decision making. We know, but at the same time pretend we don't. It's so much easier to carry on with the status quo.

The book I've set aside to finish reading on the plane is Margaret Atwood's *The Year of the Flood*. Even though it's fiction, a passage jumps out at me that encapsulates how I'm feeling. One of Atwood's characters, Toby, reflects on her past as she endures a dystopic future:

> *Surely I was an optimistic person back then, she thinks. Back there. I woke up whistling. I knew there were things wrong in the world, they were referred to, I'd seen them in the onscreen news. But the wrong things were wrong somewhere else.*
>
> *By the time she'd reached college, the wrongness had moved closer. She remembers the oppressive sensation, like the waiting all the time . . . Everybody knew. Nobody admitted to knowing. If other people began to discuss it, you tuned them out, because what they were saying was both so obvious and so unthinkable.*
>
> *We're using up the Earth. It's almost gone. You can't live with such fears and keep whistling . . .*

Before I flick the book closed, I see Atwood's acknowledgments at the end. One line catches my eye. '*The Year of the Flood* is fiction,' she writes, 'but the general tendencies and many of the details in it are alarmingly close to fact.'

Chapter Ten

THIRTEEN MINUTES
PAST MIDNIGHT

IT'S THIRTEEN MINUTES PAST MIDNIGHT. WE'RE ON YET ANOTHER plane, this time from San Diego to Boston via Chicago. All around me people are sleeping. The little boy next to me snuggles into his mother's lap as she quietly tells him a bedtime story to get him to close his eyes. I sip badly made American airline tea from a styrofoam cup and type on my laptop.

We're only halfway through our journey, but I'm already pushing myself to my limits. Physically, the travel is gruelling. The routine of packing, flying, filming, packing, flying and filming repeats over and over. Every day I stumble around zombie-like, with only the minimum hours of sleep required to function.

Mentally, the science is challenging. It is almost like I can feel my brain physically expanding. But the hardest part by far is the emotional journey this trip is taking me on. As the science gets even clearer, my tolerance for games decreases. And there are a lot of games and half-truths in the world of those who don't accept the science. So it's getting harder and harder to stomach meeting people who vehemently deny basic physics. It's as if they're oblivious to the mess they're leaving the rest of the world, especially my generation who will inherit the enormous cost in human lives and environmental damage.

Doing this work is a battle. But I don't feel cut out to be a soldier. I'm too soft; I get hurt easily; I take things personally. But then again, climate change *is* personal to young climate activists—in the same way that women getting the vote was personal to the suffragettes. The outcome of the decisions made about climate change will directly affect our futures. The level of cuts in carbon pollution we make today determines the kind of world we inhabit as we grow older, and the world our children will be born into.

People always shrug and say, 'It's politics, just get a thicker skin', and I know I need to, but I believe that talking about climate change science was never *supposed* to be about politics. In Australia, the rules of the game were set when we lost bipartisan support for action on climate change. This was in part due to Nick's actions in toppling Malcolm Turnbull and replacing him with Tony Abbott. The terms of the argument—partisan, political, ugly— weren't set by me, but I have to participate within them.

Half the space in one of my two bags is taken up with climate science books. They're not light reading! In one of them, *Six Degrees: Our Future on a Hotter Planet,* the author Mark Lynas summarises the scientific consensus for what the world will look like under each additional degree of warming, with six chapters summarising one to six degrees of climate change. It had me sobbing halfway through chapter two. Yet I can't turn away. This is my life, my future being written about.

Like those who flip straight to the star signs in the Sunday papers, I too am immensely curious—and scared—about what my future world will be like. I'm about to get married, and trying to decide, given the evidence of what's to come, whether it would be responsible to bring a child into this world. I want a family. Yet now that I know what I do about climate change, now that I've listened to the scientists and their warnings, what should I do? I understand that a baby born into the world at this point in history doesn't have great odds for the kind of happy, easy life I've been able to live up until now. Many countries will be worse off than Australia, but we're still one of the developed countries most vulnerable to the impacts of climate change.

My mind wanders briefly to the question of whether I should try to get citizenship somewhere else; somewhere safer from the worst of the climate change impacts. Then I wonder if I am not being too melodramatic. I wish I were. But if the medium- to worst-case scenarios play out, even those predicted by the conservative IPCC, there will be few places unscarred by the coming disasters. I suspect that citizenship and the legal protection it entails will hold less value in a world defined by conflict over scarce resources.

One of the other books in my suitcase is called *Climate Wars: The Fight for Survival as the World Overheats*. It was written by geopolitical analyst Gwynne Dyer after he realised that many of the world's militaries were already plotting the security implications of climate change, looking at 'scenarios that started with the scientific predictions about rising temperatures, falling crop yields and other physical effects', and then examining the political and strategic consequences.

The scenarios predicted 'failed states proliferating because governments couldn't feed their people; waves of climate refugees washing up against the borders of more fortunate countries; and even wars between countries that share rivers.' As Dyer explains in a 2008 article:

Food is the key issue, and the world's food supply is already very tight ... A 1°C rise in average global temperature will take a major bite out of food production in almost all countries that are closer to the equator than to the poles, and that includes almost all of the planet's 'breadbaskets'.

So the international grain market will wither for lack of supplies. Countries that can no longer feed their people will not be able to buy their way out of trouble by importing grain from elsewhere, even if they have the money. Starving refugees will flood across borders, whole nations will collapse into anarchy—and some countries may make a grab for their neighbours' land or water.

These are scenarios that the Pentagon and other military planning staffs are examining now. They could start to come true as little as 15 or 20 years down the road.

•

Earlier in the plane ride, Nick had handed me the September 2011 issue of the right-wing Australian magazine *Quadrant*. He suggested I read an article entitled 'How Green is their Energy' by Geoffrey Luck. Luck points out that renewable energy infrastructure (solar plants, wind farms) take up more land space than coal-fired or nuclear power stations. He claims that, therefore, renewable energy has a larger and more damaging environmental impact than coal. I can barely believe this is published as a serious intellectual argument, especially in an Australian magazine. Australia is not short on land, and Luck is measuring environmental impact simply by the amount of land used rather than the amount of carbon pollution generated. At first I think it's an *Onion*-style joke article. But as I read on, I realise he's serious. He writes:

> *Renewables must be understood in terms of the watts of energy per square metre that each source can produce . . . A biomass power plant requires about 2500 square kilometres of prime farmland to equal the output of a single 1000 mW nuclear power plant that would occupy a few hectares. Photovoltaic farms would carpet 150 square kilometres of countryside to equal the nuclear plant, while windmills would need to cover 800 square kilometres in very favourable situations.*

Luck's argument is based on a pre-1800s understanding of the environment as the landscape—nothing more than physical space on the surface of the earth. He completely ignores modern science's revelation that the environment includes, and is dependent on, the health of the atmosphere. What we do to the air has major implications for the land, ocean, animals and human health. The carbon pollution generated from the power plants Luck finds so benign is far more environmentally damaging than the extra square metres taken up by solar panels and wind farms. Luck also ignores the fact that much renewable infrastructure can be generated away from arable land—like in deserts or even in the oceans, with tidal power or offshore wind turbines.

I flip through the rest of the magazine Nick gave me and come across another curious article. It's by the president of the Czech Republic, Vàclav Klaus, and is entitled 'Climate Change: The Dangerous Faith'. The language in it sounds familiar. In 2009, during a *Four Corners* interview, Nick had said that climate change had become 'the new religion' for the 'extreme left'. I wonder if he and Klaus share talking points. The article is agonising to read, given our recent discussion with Professor Naomi Oreskes about the anti-communist Cold War physicists who were the first to campaign against climate science. Klaus rails against environmentalists and others who have accepted the 'belief' of climate change. He writes:

> *They want to change us, to change our behaviour, our way of life, our values and preferences, they want to restrict our freedom because they themselves believe they know what is good for us. They are not interested in climate. They misuse the climate in their goal to restrict our freedom . . . They don't care about resources, poverty or pollution—they hate us, the humans, they consider us selfish and sinful creatures who must be controlled by them. I used to live in a similar world—called communism . . .*

I wish Nick would take me to meet Klaus so I could assure him I don't hate humans. What motivates me in my climate change activism is not hate, but love. When I decided to set up the Australian Youth Climate Coalition, it was because I felt so lucky to be alive in a world that sustained such a dizzying array of life. I love humanity, all its diversity, and the countless other species that we share this planet with. The thought that it could all be destroyed—not just polar bears and pandas and turtles, ancient forests and farmland, and glaciers and reefs, but most importantly to me, people's lives—is earth-shatteringly awful. When I learned about the suffering that communities were *already* experiencing from climate change, it was a sense of moral outrage that spurred me to act.

The other young people leading the youth climate movement in Australia have unfailingly similar motivations. Ask my friend Sam from Yackandandah in rural Victoria, who fought the Black Saturday bushfires alongside his family and neighbours. He loves this country and wants to see it saved from the ravages of climate change–induced extreme weather. Ask AYCC's current national director, Ellen. She grew up in Mildura along the banks of the Murray and has seen the river she loves slowly deteriorate over the course of her lifetime. Ask my old co-director at AYCC, Amanda, and she'll tell you how she felt while reading Tim Flannery's book *The Weather Makers* on a trip to the ancient Tasmanian forests. Amanda couldn't bear the thought of thousand-year-old trees and the animals that depend on them being destroyed by climate change.

These stories are important, yet they are only three examples in a world that is already chalking up thousands of victims of climate change—humans, animals and ecosystems. So far the burning of fossil fuels has raised the earth's temperature by a global average of only 0.8 degrees. This has *already* destroyed the lives of thousands of human beings around the world. The aid agency Oxfam Australia states on its website that

> although no single climate event can be attributed directly to climate change, climate modelling has predicted a range of climate catastrophes from a greater risk of heat waves and wildfires in Europe to flooding in southern Asia. Tragically this modelling has been remarkably accurate, with Russia and Pakistan experiencing these exact conditions.

Climate-related disasters in 2010, according to the global insurer Munich Re, have caused US$130 billion in losses, rising food prices and 21,000 deaths in the first three-quarters of the year.

Author Bill McKibben has catalogued the ways in which our old familiar planet is melting, drying, acidifying, flooding and burning in ways humans have never seen. We've created a new planet, he argues. It's still recognisable but fundamentally altered, to the extent that 'it's a different place. A different planet. It needs a new name'. McKibben uses his book *Eaarth* to take stock of what this new planet we live on means for humanity, now that we've warmed 0.8 degrees Celsius as a global average. He starts with the implications of hotter oceans, which create more storm clouds and therefore lead to more rain and hail:

> *Total global rainfall is now increasing 1.5 percent a decade. Larger storms over land now create more lightning, according to the climate scientist Amanad Staudt. In just one day in June 2008, lightning sparked 1,700 different fires across California, burning a million acres and setting a new state record. 'We are in the mega-fire era,' said Ken Frederick, a spokesperson for the federal government . . .*
>
> *By the end of 2008 hydrologists in the United States were predicting that the drought across the American southwest had become a 'permanent condition' . . . researchers calculate that the new aridity and heat have led to reductions in wheat, corn, and barley yields of about 40 million tons a year. The dryness keeps spreading. In early 2009 drought wracked northern China, the country's main wheat belt. Rain didn't fall for more than a hundred days, a modern record. The news was much the same in India, in southern Brazil, and in Argentina, where wheat production in 2009 was the lowest in twenty years . . .*
>
> *From the flatlands to the highest peaks . . . In the spring of 2009, researchers arriving in Bolivia found that the eighteen-thousand-year-old Chacaltaya Glacier is 'gone, completely melted away as of some sad, undetermined moment early this year' . . . These glaciers are the reservoirs for entire continents, watering the billions of people who have settled downstream precisely because they guaranteed a*

steady supply. 'When the glaciers are gone, they are gone. What does a place like Lima do?' asked Tim Barnett, a climate scientist at Scripps Oceanographic Institute. 'In northwest China there are 300 million people relying on snowmelt for water supply. There's no way to replace it until the next ice age.'

McKibben's book is filled with these statistics and hundreds more, but he urges us not to let our eyes glaze over. 'These should come as body blows, as mortar barrages, as sickening thuds,' he writes. 'The Holocene [our current geological period] is staggered, the only world that humans have ever known is suddenly reeling.'

This is why aid agencies such as World Vision, Oxfam, UNICEF and the World Food Program are now just as deeply involved in climate change campaigning as traditional 'environment' organisations. They understand that climate change is literally a matter of life and death for people already living in poverty.

So to people like President Klaus, who says 'they hate us, the humans', I say this: you are unequivocally wrong. Most people working to solve climate change care deeply about humans, and human suffering inflicted by any cause, whether it be extreme poverty, violence and conflict or discrimination and disease. I feel crushed to live in a world where there is already so much suffering, and horrified at the way it is increasing due to climate change. As American journalist Ross Gelbspan writes:

An acknowledgement of the reality of escalating climate change plays havoc with one's sense of future. It is almost as though a lone ocean voyager were suddenly to lose sight of the North Star. It deprives one of an inner sense of navigation. To live without at least an open-ended sense of future (even if it's not an optimistic one) is to open one's self to a morass of conflicting impulses—from the anticipated thrill of a reckless plunge into hedonism to a profoundly demoralizing sense of hopelessness and a feeling that a lifelong guiding sense of purpose has suddenly evaporated.

This slow-motion collapse of the planet leaves us with the bitterest kind of awakening. For parents of young children, it provokes the most intimate kind of despair. For people whose happiness derives from a fulfilling sense of achievement in their work, this realization feels like a sudden, violent mugging. For those who feel a debt to all those past generations who worked so hard to create this civilization we have enjoyed, it feels like the ultimate trashing of history and tradition. For anyone anywhere who truly absorbs this reality and all that it implies, this realization leads into the deepest center of grief.

Perhaps this is one reason that some people refuse to read articles about climate change. Most Australians are deeply compassionate. We respond generously to help others, whether it be in the wake of the Queensland floods, the Victorian bushfires, or the Japanese tsunami. During Christmas in Australia, demand for volunteer roles like serving in soup kitchens often outstrips the need. But people resent being forced to care about something. Maybe good people fear that when they open the floodgates and get their heads around what's happening to the climate, it will overwhelm them. Maybe this is why so many people put climate change in the 'I'll deal with it later' part of their brains.

Chapter Eleven

THE NUTTY PROFESSOR: BOSTON

ONE OF MY TWO BAGS MAKES IT TO BOSTON. THE OTHER—THE ONE with my carefully chosen outfit for our next meeting—is stranded in either San Diego or Chicago, where Max had filmed Nick and me in an impromptu argument about earth's climate sensitivity at the airport Starbucks. Now it's two in the morning and there's nothing I can do except pray it turns up tomorrow.

My prayers are answered and my bag arrives at the hotel the next day. As suspected, it was enjoying a night in Chicago. I reunite joyously with my knee-length black dress and a particularly nice pair of glossy black heels. I put them on, and soon we're on our way to meet Professor Richard Lindzen at his home in Boston.

Professor Lindzen is one of the highest profile climate sceptics and the high priest of the 'no feedbacks' brigade. This is partly because he's one of a tiny handful of climate scientists who disagree with the reports of the Intergovernmental Panel on Climate Change, despite having contributed to them.

Lindzen is a professor of meteorology at the Massachusetts Institute of Technology and a member of the National Academy of Sciences. However, he is also a 'contributing expert' to two right-wing, libertarian, anti–climate science think tanks: the George C Marshall Institute and the Cato Institute. Lindzen used to be a highly respected atmospheric scientist but his reputation has taken a battering over the past decade. The American journalist Ross Gelbspan reported in 1995 that Lindzen 'charges oil and coal interests $2,500 a day for his consulting services; his 1991 trip to testify before a Senate committee was paid for by Western Fuels, and a speech he wrote, entitled "Global Warming: the Origin and Nature of Alleged Scientific Consensus" was underwritten by OPEC [Organization of Petroleum Exporting Countries]'.

Despite this, Nick is still one of Lindzen's biggest fans. Nick often prefaces his rebuttals to climate science with the words, 'The Dick Lindzens of the world say . . .' So I know that Lindzen is a Big Deal to Nick. This meeting will be the intellectual pinnacle of Nick's argument that climate change isn't anything to be worried about.

I have no doubt that it will be the hardest point of the journey for me. Dick Lindzen is as good as climate sceptic spokespeople get. He'll have a point prepared for every question I bring up. I'm also not relishing telling a successful 71-year-old scientist that his theory is utterly debunked. Even though the vast majority of scientists agree with me, it still feels impolite. But Eleanor Roosevelt's words—'Do one thing every day that scares you'—pop into my head. The stakes are very high, and I feel a bit like Harry Potter before his final showdown with Voldemort.

Nick, Naomi and I had briefly talked about Lindzen's cloud theory yesterday. 'Here's the thing about Dick Lindzen,' Naomi said. 'Back in the 1980s he raised the question of cloud feedbacks. We all know that clouds can make things cooler. It's cooler here right now than if it weren't a cloudy day—we all know that. That's common sense and it's consistent with our experience and it's consistent with physics.' Yes, I thought, of course clouds stop heat reaching the earth.

Naomi continued. 'So back in the early 1980s, [Lindzen] said look, if you increase the amount of carbon dioxide in the atmosphere it will warm the planet, he agreed about that. He didn't disagree . . . *but,* he said, it will cause an increase in the hydrologic cycle—so he agrees with that too—it will cause more evaporation, more water vapour in the atmosphere and that increased water vapour will create more clouds, and that will be a negative feedback.' Right, I thought. So Lindzen agrees that warming oceans will release water vapour, but he believes that water vapour will change into cloud form instead of remaining a heat-trapping greenhouse gas.

Climate science tells us that increased water vapour absorbs more heat and is therefore a positive feedback mechanism. But Lindzen believes that instead of water vapour staying in water vapour form (where it absorbs heat) it will form cloud cover that blocks out the sun's warming rays, and slows warming. 'That's why he comes up with this "Only 1 degree [of warming]" even though most scientists think it's substantially more than that,' said Naomi.

So Professor Lindzen argues that a doubling of CO_2 will cause the global average temperature to rise by only 1 degree. This is outside the range of between 2 and 4.5 degrees that the IPCC says is most likely.

'So in 1980 that was a very important point to have raised,' Naomi said. 'He was totally right to raise it . . . It got published in the leading scientific journals in the world and people said, you know, Dick Lindzen has raised an important point.' What happened next? As happens in science, people tested the theory against reality. 'For thirty years scientists have studied cloud feedbacks in excruciating detail,' said Naomi. 'So here we are thirty years later and we now have direct observations. And here's the thing: the cloud feedbacks have *not* prevented warming.' In fact, scientists now know that certain types of clouds exacerbate warming by absorbing heat radiation emitted by the earth's surface.

Naomi's point is echoed by all the articles about Lindzen's theory I've read in preparation for this meeting. Lindzen argued that for a doubling (a hundred per cent increase) of CO_2, we would only get 1 degree warming, because clouds would act to block the sun from heating the earth. Yet here we are with a thirty-eight per cent increase in CO_2 and the earth has *already* warmed by 0.8 degrees. The *observational evidence* alone has proved his theory wrong.

This seems reasonably clear to me, even as a non-scientist. So why hasn't Lindzen changed his tune? Nick offered one possible explanation yesterday: 'I guess that observational evidence is open to interpretation, is it not?'

If Nick is about to argue that the surface temperature records compiled by NASA and the other scientific agencies are wrong, he's on shaky ground. I thought Professor Muller had convinced him back in San Francisco that the temperature record was solid.

Naomi Oreskes compared Lindzen to Harold Jeffreys. Jeffreys was a brilliant geophysicist who held the same academic chair at Cambridge that Sir Isaac Newton had. 'This is one hell of a smart guy,' Naomi said. 'But he went to his death without accepting plate tectonics [the theory that explains continental drift; it's one of the pillars of geosciences]. So it's also important to realise people can be really intelligent but they can get stuck,' said Naomi. 'I think Dick Lindzen is kind of a stuck guy. And the smarter someone is I think the more at risk they are of that.'

•

We drive through Boston's picture-perfect outer suburbs. Grand old houses sit behind white picket fences among manicured lawns and flowering shrubs. American flags wave in the breeze. Even the leaves look perfect: just the right shade of green.

Lindzen's house is just past a small lake. Recreational fishers scatter along its edge, enjoying the idyllic weather. We pull into the driveway of a blue house with a sign out the front supporting a candidate in the local school board election. Nick jumps out of the van enthusiastically, eager to meet Lindzen. I can almost feel his excitement.

I linger inside the van, reading and rereading the notes I've made for today. I'm nervous. I try to make sure I completely understand Lindzen's argument and am able to rebut it coherently. By now, I've spent more than a few spare hours studying clouds to make sure I can do my best in this intellectual cage match. Eventually the van driver needs to leave for another job. He senses my reluctance to go inside the house. 'Good luck. You'll be great,' he says, smiling encouragingly.

I watch the van drive away and then walk slowly up Lindzen's front steps. The door is open. Lindzen is talking to Nick a few metres away. He's short, with thick glasses and a beard. I hover in the doorway and smile at him in that tentative way you do when you want to get invited into a conversation—or in my case, a house. He's facing me so I'm expecting a 'Hi, come on in' at the very least. Instead, Lindzen ignores me. He's not interested in small talk before the interview, or even acknowledging my presence.

Cautiously and slowly, I start to walk in anyway. But then I halt. Stale cigarette smoke hits me like a wave. The stench so overpowering that I start coughing and have to retreat back to the front steps. I'd read the profiles of Lindzen that mentioned his chain-smoking, of course, but I hadn't imagined this challenge when preparing for our meeting.

Max is tossing up whether to film the meeting inside, in the lounge room, or out on the deck. I beg Max to make it outside, for the sake of everyone's lungs. But he decides the lounge room better reflects Lindzen's character. I steel myself for a few hours of shallow breathing and sit in the sun on the front steps, running over the key points in my mind.

Eventually Max directs me to go inside. We all walk up a narrow staircase leading to Lindzen's den. In this small room—almost an attic—on the third floor of his house, Lindzen's ham radio equipment is set up. Pete and Max film Lindzen as he uses the radio: a big black box covered in knobs and dials hooked up to a computer monitor. Nick and I hover behind Lindzen and watch as he uses it to speak to someone on the other side of the world. The room is small, hot and the windows are closed. The cigarette stench is even worse up here, courtesy of a green and white ceramic ashtray overflowing with cigarette stubs that rests next to Lindzen's right elbow. I'm finding it increasingly difficult to breathe while Lindzen scans the radio channels.

I distract myself by thinking about how this ham radio scene will play with the television audience. What does mainstream Australia think of ham radio users? I have no idea. To me, ham radio has undertones of survivalism. I almost expect to hear gruff military voices speaking in half-coded language about an alien invasion they're busy covering up. When communications go down in one of those apocalypses that are supposed to be around the corner, only ham radio users will retain the tools of communication.

I do a quick Google search later that evening for 'ham radio' plus 'survivalist'. Returning 264,000 results, one of the first sites that appears is *The Zombie Hunter: A Survivalist's Journal: The Family Man's Guide to Surviving the Zombie Apocalypse.* The author identifies himself as The Zombie Hunter and writes: 'To the survivalist, amateur or "ham" radio will always be a relevant hobby . . . This will be a crucial skill for survival when the crisis comes. And I cannot stress how important the role of amateur radio operators will play in the long-term struggle against the hordes of undead.'

I'm not suggesting Lindzen is anything like Mr Zombie Hunter. Lindzen says he got involved in ham radio from a young age and simply finds it an interesting hobby. And I may disagree with his views, but I don't think he's a nutter. He was once a highly respected atmospheric scientist.

Now that he's acknowledged my existence, I get the chance to examine him more closely. His jacket is stained. He speaks slowly and with narrowed eyes. His house looks new and he tells us that much of it had to be rebuilt after a fire that caused major smoke damage to the walls.

After Max and Pete have enough ham radio footage, we all file back down the narrow staircase to the living room. Pete has set up some bright studio lights to illuminate our faces in the dark room. My hands shake as I sit and my heart is beating fast. This is definitely going to be hard, but I'm determined to bring everything I've got to the conversation.

We start with a bang. Lindzen says, 'It is true almost all people working in this field accept that the last century and a half have seen a fraction of a degree.' *A fraction of a degree?* So this is how Lindzen is going to try to defend his theory? He's going to argue that the world *hasn't* warmed by almost 1 degree Celsius with only a thirty-eight per cent increase in greenhouse emissions. To acknowledge this would immediately disprove his theory that the world will only warm by 1 degree Celsius with a hundred per cent increase in greenhouse emissions.

'Three-quarters of a degree,' I say, surprising even myself with my willingness to correct Lindzen so early into our conversation. He seems startled too. 'Say that again. Three-quarters of a degree. Why do you say that?' he asks. 'My understanding is that that's what the temperature records show from NASA and NOAA,' I say. This is all pretty basic stuff. Is he *really* going to dispute the temperature record?

'How are they taken?' asks Lindzen, referring to the temperature records. 'Through weather stations,' I reply, and start to tell him about meeting Muller. Lindzen pulls out what he clearly thinks is his trump card: 'There is an error bar.'

'When somebody says three-quarters of a degree, OK, you are thinking they have a thermometer that reads that carefully,' says Lindzen. 'What they have are hundreds of thermometers reading all over the place and they look at how they deviate and it's a cloud of data scattered all over the place and they average that and they get almost nothing and then they blow up that scale to three-quarters of a degree and then if you read the document it says plus or minus, which means it could have been three-quarters of a degree, it could have been 0.4, it could have been 1 . . . It's not a precise number.'

OK, so he *is* disputing the temperature record, I think. He adds: 'The one thing you know about it, that you've just said, if you thought about it a moment, [is that] it's small.'

I don't bother arguing the point he's insinuating about NASA's temperature records being unclear. I think the TV audience will be smart enough to compare this with Professor Muller's analysis showing NASA's records are actually fine. But Lindzen's last point— that a temperature increase of 0.8 degrees is 'small'— is another matter. 'It has already shown big impacts on the climate,' I say to Lindzen.

The fact that what sounds like a small increase in global average temperature actually has enormous impacts is the whole reason why anyone worries about climate change. In the course of a day in Sydney, temperatures frequently rise by six degrees or more. But, as Elizabeth Kolbert writes in her book *Field Notes from a Catastrophe*, 'Average global temperatures have almost nothing to do with ordinary life. This is perhaps best illustrated by the ups and downs of climate history.' The largest temperature changes of the past million years are the glacial cycles. The difference in global average temperature between the last ice age and the climate in which human civilisation developed is only around 5 degrees Celsius, plus or minus one degree (although some scientists say the range is broader, between 4 and 7 degrees Celsius). Only 5 degrees, and this difference caused vast ice sheets to cover much of North America, northern Europe and Asia. So much water was frozen that sea levels were low enough for a land bridge to connect Siberia and Alaska.

Lindzen is an atmospheric scientist, so he knows the difference between global average temperature and daily weather changes as much as any climate scientist. But he continues down this line: 'No, it's small,' he says. 'For instance, how much does the temperature change between day and night in Sydney?'

'I think that's kind of confusing the point,' I say. 'Because we're talking about global average temperature, not day-to-day weather. It's much warmer at the poles, whereas the tropics aren't experiencing that much warming – but the *global average* is about three-quarters of a degree according to NASA.'

Lindzen refuses to accept this. 'There's been a *small* change in the global mean temperature,' he says again. 'Let's start with that.' I sigh. Let's move on.

Next, we argue about whether or not humans are responsible for this warming. When we met Matthew England in Hawaii, he'd explained how climate scientists know that the additional greenhouse emissions in the atmosphere post–Industrial Revolution come from human activities. First, Matthew explained that countries report their emission levels. Various international agencies, including the UN Environment Program, the International Energy Agency and the UN Secretariat (which monitors compliance with the Kyoto Protocol) keep data sets of national greenhouse pollution levels. Second, scientists can track the isotopic signature of the type of CO_2 in the atmosphere. Ancient carbon (from fossil fuels) has a different isotopic signature to new carbon from natural processes like volcanoes.

Yet Lindzen's next point is to dispute the fact that the extra greenhouse gases come from humans. 'You know, Murry Salby [a professor of environmental science at Macquarie University] in Australia has pointed out something,' says Lindzen. 'And he's correct in one sense. I don't know if he's right, ultimately, but there has been a study . . . they assume that only man's emissions would have a certain isotopic signature. He pointed out there are plenty of natural processes that have the same signature. So it would be impossible to pin it down.'

'My understanding was that countries publish the amount of emissions that they actually emit,' I say. 'And that there's more than enough to account for the CO_2 increases.'

'Absolutely,' replies Lindzen, 'but that doesn't show that it *is* what accounted for it.'

I can't believe this line of argument. 'What else would you say [caused the warming]?' I ask.

'Anything else that produces the same isotopic signature,' replies Lindzen.

By now, my mind is boggling. We have a certain amount of greenhouse gases of a particular isotopic signature in the atmosphere. This matches the amount of greenhouse gases that the nations of the world emit, which is tracked and reported each year. Yet Lindzen is arguing that something other than humans might have emitted the greenhouse gases that have caused the planet to warm.

We are never going to agree on this, so we move on. Lindzen attacks climate models. He argues that they predicted we would have seen more warming than we have seen. I'd expected Lindzen to bring this up. But I also know he's talking about models that don't include the temporary cooling effect of aerosols (remember Ryan's lydar back at Mauna Loa?). He's also not counting the time it takes for oceans to absorb heat. When the climate models input these factors, the planet has warmed almost exactly as much as they predicted.

Figure 11.1 is an example of the results generated from the model used in the IPCC's fourth assessment report compared with observations.

Figure 11.1 IPCC model results compared with observations

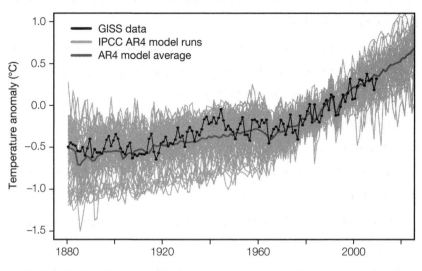

Source: Plotted by Grant Foster

Sometimes the models have even been *too* conservative. Actual climate change impacts, such as sea level rise, have in some cases tracked at the upper end of predictions from climate models.

Lindzen knows the models track well against the observations, so he makes a pre-emptive strike. 'Now the way the models hide it is by saying well there's aerosols, that cancels some of it, that's why we haven't seen it,' he says. 'And then you look at each model. Each model uses a different value for aerosols. You ask the aerosol people and they say we don't know what the value is. So it's become a fudge factor. We're now in this position where there is a knob that you can make any model agree with any data.'

Lindzen is ignoring the fact that we know about feedback mechanisms from the ice core data, not just the models. 'But the information that scientists are telling us about climate sensitivity doesn't just come from models, it comes from ancient climate history,' I say.

'My understanding is that the paleoclimate community thinks it's quite bizarre that you're advocating a 1 degree climate sensitivity when they understand the glacial cycles and can learn that there's 3 degree sensitivity.'

I've cut to the heart of Lindzen's argument about climate sensitivity. Climate scientists use information from the past to predict the future of our climate. And in this data from the past, they find clear evidence of positive feedback mechanisms that accelerate warming. Lindzen, however, argues positive feedbacks will not worsen the initial warming from CO_2. Instead, he believes a *negative* feedback mechanism—clouds—will kick in to prevent warming.

Lindzen seems angry that I've brought up the ice cores. 'That's nonsense,' he says. 'The paleoclimate people have usually never worked with theory of climate. The papers by Jerry Robe, by Peter Huybers and Milankovic are theory . . . It's pure ignorance to say paleo demands it.' As far as I know, this is a very solid part of climate science that rests on the work of a lot more than a handful of papers. Lindzen continues: '[James] Hansen has said this for years. It doesn't make any sense and I don't think it's even widely accepted. The paleo community has grabbed on to it, says oh, now we're relevant, but I mean I don't think it matters more than that.'

We're going to have to agree to disagree. Paleoclimatology is the study of how the climate changed in the past. This seems pretty relevant to the study of how the climate will respond today. And I doubt very much that an entire field of science would make up something so important just to be 'relevant'.

Lindzen wants to move on. To reach problematic climate change scenarios, Lindzen says, 'You'd need to have a strong, positive feedback, so strong that if you regarded the earth as a system, no engineer would ever build it this way.'

Essentially, Lindzen is arguing that climate change can't be real because if it were, the earth would have the capacity to self-destruct through feedback mechanisms that amplify warming. This argument has almost religious ring to it: God wouldn't have built an earth with the ability to implode like this, therefore it can't be happening. 'The earth has lasted 4.5 billion years,' he says. 'There are plenty of ways it could have gone over the top, but it hasn't.'

It's a nice, comforting thought. Unfortunately, the ice ages prove that the earth *does* have the ability to shift between extreme climate states when triggered by external forcing. In ancient history that trigger was the sun or changes in the earth's orbit; now the trigger is humans.

This is the background to Lindzen's theory about clouds, which he sees as the earth's natural defence mechanism against climate change. 'Clouds are the only reason these models have a large feedback, positive feedback,' he says. 'Water vapour gives it a factor of 2 but water vapour only acts where you don't have clouds.'

'So you mean that for every degree of warming, with water vapour there's another 2 degrees [warming],' I say. Lindzen just explained succinctly how positive feedbacks worked.

'No, that's baloney,' he says, crushing my sudden glimmer of hope that he'd changed his mind. 'That's purely hypothetical. It says if relative humidity stays fixed and the clouds don't do anything and the water vapour covers everything you will get that.'

But Lindzen's cloud theory was proven wrong, I try to argue. We go around in circles talking about climate sensitivity. I emphasise the data from the ice cores and the models showing climate sensitivity to be around 3 degrees. Lindzen argues the models are wrong and there's no empirical evidence for positive feedbacks.

The rest of the discussion is just as exasperating. Lindzen blithely argues that if climate change is happening, humans can just adapt. 'When people retire in North America they usually move to places that are a different temperature by about 10, 15 degrees [Farenheit compared] with where they move from,' he says. 'Clearly at the human level we adapt to that with alacrity. You would be surprised how few people here retire to Alaska,' he says.

But then Lindzen makes a mistake. He tries to quote a former NASA employee, Joanne Simpson. 'You had a former director of research, Joanne Simpson, who retired a few years ago, she's now dead,' says Lindzen. '[She] said, "Now that I'm retired I can admit that I don't think this is correct."'

I jump on this immediately. 'You can't just cherry pick her quotes,' I say. I have seen the full quote, and there's a whole lot more that Lindzen didn't mention. Here are Joanne's full words:

> Since I am no longer affiliated with any organization nor receiving any funding, I can speak quite frankly. What should we as a nation do? Decisions have to be made on incomplete information. In this case, we must act on the recommendations of Gore and the IPCC because if we do not reduce emissions of greenhouse gases and the climate models are right, the planet as we know it will in this century become unsustainable. But as a scientist I remain skeptical.

Lindzen tries to backtrack. He claims that Joanne Simpson expressed doubt about the science at a meeting she and Lindzen both attended, *without* adding that we should act any way. Then, despite being the one who brought up her quote, Lindzen says: 'You can quote scientists as much as you want. Science, as I've often said, is not a source of authority. It's a source—it's a methodology for inquiry.'

The conversation degenerates from there. Lindzen probably gets as frustrated with me as I do with him. Towards the end, when it comes down to a matter of which scientists to trust, I bring up the role of outliers in science and public policy. There will always be contrarians who disagree with the mainstream peer-reviewed science, but what weight should we give these individuals in important public policy decisions? I use Lindzen's scepticism of the link between second-hand smoke and cancer as an example. Unlike the majority of the medical and scientific community, Lindzen refutes that there is a link.

'I have argued,' says Lindzen, 'as [have] most people who have looked at it, that the case for second-hand tobacco [causing harm] is not very good. I'm not worried about that.' I don't ask him what his view is on direct smoking. Surely he can't dispute that one—but he tells me anyway. 'With first-hand smoke it's a more interesting issue,' he says. 'There's clearly an issue and, you know, one looks at the statistics and it was done by meta-analysis. That's usually the case when you don't have great studies and you have to combine them. The case for lung cancer is very good—but it also ignores the fact that there are differences in people's susceptibilities, which the Japanese studies have pointed to.'

I'm confused. Is Lindzen is saying there's an 'issue' with smoking, but that because studies were combined, they might not be accurate? 'I'm simply saying anything in science requires you look at it,' says Lindzen. 'You know, I have always found it profoundly offensive that to question something indicates you're doing something wrong. It pays to look at the studies.'

'Have you read the studies?' he asks, referring to passive smoking. I reply with my own question: 'It seems to me that the evidence for passive smoking leading to health problems is quite clear, so if you're questioning that, why should I trust you on climate science?'

Then Nick finally enters the fray. 'What's this got to do with climate change?' he asks irritably.

'Well I think it has a lot [to do with it] because it comes down to who you trust on the issue of science,' I say.

Nick replies: 'There is a debate about passive smoking and its impacts on health. We know that. We know there's a debate about that. Who's suggesting that debate is over? And I really don't see the relevance to a debate about, you know, the extent to which we should be concerned about CO_2 emissions.'

Nick then changes the topic and our conversation begins to wrap up. Finally, I ask Lindzen the same question I asked Joanne Codling and David Evans in Perth. 'What if you're wrong?'

'That's a common question,' replies Lindzen. 'And, you know, it doesn't bother me much, for a very simple reason. Almost every political response [to climate change] does nothing, whether I'm wrong or right.' I realise it's all a game to Lindzen, because he's a pessimist about the world's ability to cut carbon pollution anyway. 'The only thing it does is reduce the robustness of society to deal with change which will occur whether it's due to man or not,' he says. 'Any time you cut prosperity and wealth you increase vulnerability of a society.'

The meeting ends. It's been over two hours since we arrived. I tell Lindzen we'll have to agree to disagree. His wife walks out from the kitchen, says hello, and kindly gives us all bottles of water. And then we leave.

Outside, I gulp in big breaths of clean air. It feels great. Max wants me and Nick to stroll around the block, debriefing while the meeting with Lindzen is still fresh in our minds. Max and Pete walk in front of us, filming while we talk.

'How do you think that went, guys?' asks Max. 'Anna?'

I'm still reeling from the fact that Nick and Lindzen both admitted, on camera, that they don't think the link between passive smoking and health problems is settled. 'Well, he's obviously a smart person,' I say, 'but it doesn't mean he's smart about everything. I was really surprised that he brought up that he doesn't believe that passive smoking causes health problems—because it clearly does.'

At this, Nick loses his temper. I've suspected this has been coming ever since our conversation with Naomi. His face reddens and his voice gets louder. 'Clearly there is an ongoing debate,' he says. 'I know that, everybody knows, there's an ongoing debate. No one says it's been absolutely established beyond doubt that passive smoking kills people.'

'But we still think it's proven enough to act,' I say.

'It's entirely irrelevant to this debate,' continues Nick, 'and one thing that really pisses people like me off is this red herring that people like you raise about tobacco. It's outrageous and you should stick to the debate . . . I think is outrageous and you tried to slander him with this stupid claim about tobacco. I got very upset by that.'

I try to explain why it's important to look at the credibility of someone's scientific record before accepting their claims as fact. 'I do believe it's relevant to look at the past . . .'

Nick interrupts before I can finish my sentence. 'It's utterly irrelevant to raise issues about tobacco,' he says, 'and it is part of the slander of the environmental movement in attacking anybody who doesn't accept what they say.'

I try again. 'Nick, I do think we have to look at the past claims of people who are saying . . .'

'Rubbish,' says Nick.

He thinks I'm being unfair by bringing up Lindzen's outlier status on passive smoking. I think it's reasonable to put someone's views on one topic in context of their views on other topics. Otherwise, how do non-scientists like Nick and I determine their credibility in the scientific community? If someone is a habitual contrarian, or known for making extreme or incorrect claims, why should one take them seriously in the climate debate? And as far as I'm concerned, Lindzen's assertions in both the health and climate debates have been proven unlikely to be correct. As the website *Skeptical Science* states in an article on Lindzen:

> *There is of course nothing wrong with being occasionally mistaken in science. The problem arises when a scientist is consistently wrong and fails to learn from the corrections advanced by other scientists or by nature, especially when we're asked to believe that he is right and virtually every other scientific expert is wrong . . . Lindzen does present a mostly coherent, consistent alternative hypothesis to the anthropogenic global warming theory. There's only one problem: as discussed above, every single one of these arguments is inconsistent with the observational evidence.*

No matter how many times I try to explain this to Nick, he won't budge. 'The issue of feedback, positive or negative, to me is the absolute essence of this whole debate,' he says, 'and it's the arguments of people like Dick Lindzen that mean that I do not accept, and I am not convinced, that human emissions of CO_2 are causing dangerous global warming.'

•

It's been a tumultuous day, but that evening Nick and I have a pleasant dinner at Boston's iconic docks. We eat clam chowder, the city's specialty, as ships sail into the harbour behind us. Max films the whole thing, of course. Nick and I joke around—partly for the camera, partly because we need to let off steam after an intense few hours. Despite losing his temper earlier, Nick seems happy as we eat our chowder. I just try to pretend he's a friendly distant relative with whom I disagree on climate change. And as long as we avoid talking about climate change, we get on fine. These days, hanging around with Nick and the film crew is the extent of my social life, so I'd better make the best of it.

That night, as I drift to sleep, I replay the meeting in my mind. Lindzen was much less intimidating in real life than I'd built him up to be. From the way Nick talked about him, I'd expected bullet-proof arguments and a 'case closed' finish. Instead, I found an abrasive 71-year-old professor who, while obviously intelligent, refused to accept NASA's observational evidence. If that's the best you can come up with, I think sleepily, there's no way the audience is going to be persuaded.

Still, Nick might think the same about my arguments. His words at the end of our debrief echo in my mind. 'I do *not* accept and I am *not* convinced . . .'

I wonder how history will look back on those words.

Chapter Twelve

ATTACK DOGS:
WASHINGTON DC

OUR TRAIN FROM BOSTON FINALLY PULLS INTO UNION STATION IN Washington DC. Max films Nick and me walking through the great vaulted space. It's such an impressive building, filled with columns and arches, stone inscriptions and statues, white granite, marble and gold leaf.

It's very different from Sydney's Central station. I feel suddenly homesick, missing my fiancé Simon. Only a few months ago we'd been in this station together. He spent a week with me here while I was meeting with American environmental organisations as part of my Churchill Fellowship research.

I have a lot of friends in Washington DC. It's where the American equivalent of the Australian Youth Climate Coalition (called the Energy Action Coalition) is headquartered. After having the idea for AYCC in Montreal at the end of 2005, I'd been so inspired by the fledgling Energy Action Coalition that I'd decided not to come back to Australia straight away. Instead I'd caught the Greyhound bus to New York City, stayed in the apartment of one of their founders and soaked up all the information I could about their organisation.

Despite the distance, I'd remained friends with the Energy Action team. When I returned to the United States as an exchange student in 2007, I helped out with their No New Coal campaign and their inaugural youth climate summit (which has now grown into an annual conference attracting 10,000 young climate activists). It's an amazing feeling to be part of a global youth climate movement, with friends all over the world. Now that I am back in DC it is hard not to pick up the phone to organise a catch-up. But I know I have to keep focused on the task at hand.

The task at hand is unfortunately not pleasant. Max had tried to keep our Washington DC interviewees secret. However, neither Nick nor I were too keen on this idea. The trip was hard enough without added surprises. So unbeknown to Max and in a rare moment of camaraderie, Nick and I had swapped the names of our DC spokespeople while waiting in San Diego airport.

My pick was the chief oceanographer of the US Navy, Rear Admiral David Titley. Nick's pick was Marc Morano, the notorious Republican media personality and blogger. As soon as Nick told me this, alarm bells started ringing. I'd heard of Morano. And what I'd heard wasn't good. But I hadn't yet had the chance to read up on him properly.

At our hotel, I settle into the big blue couch in my room. I want to spend a few hours researching Morano since he'll be the first person we'll meet in the morning. After less than five minutes of Google searching my heart sinks into my stomach. I wonder what on earth Nick was thinking in selecting Morano as a spokesperson. I can't believe he would stoop so low.

A rapid-fire email exchange with one of the most respected environmental journalists in the United States confirms my fears.

Morano is one of the worst of the worst of the Republican attack dogs. I am told he plays dirty and is widely recognised as a reputation-smearing brute.

Morano made his name in the early 1990s as reporter and producer for a television talk show hosted by ultra-right wing personality Rush Limbaugh. He later moved on to work at Cybercast news service, which is owned by the Media Research Center. The Media Research Center is a conservative news organisation that aims to stamp out progressive values in news reporting and replace them with right-wing content. *The Australian* reports that the centre received US$50,000 from Exxon Mobil in 2010 as part of its campaign against climate science. After this was revealed, Exxon announced it would no longer fund the centre.

Morano was the first source to launch the initial Swift Boat smears that de-railed John Kerry's campaign to become US president. The Swift Boat smears were allegations about Kerry's military record that were later exposed as 'extremely suspect, or false'. The *New York Times* opined that the Swift Boat attacks were 'orchestrated by negative-campaign specialists deep in the heart of the Texas Republican machine'.

Morano's 'negative-campaign' talents have recently been used in his role as communications director for Republican senator James Inhofe. Inhofe is a raging climate denier. 'Catastrophic global warming is a hoax,' he claimed in 2003 on the floor of the US Senate. 'Alarmists are attempting to enact an agenda of energy suppression that is inconsistent with American values of freedom, prosperity, and environmental progress.'

After working for Inhofe, Morano turned his attention to climate denialism full time through his blog *Climate Depot*. Morano routinely publishes the email addresses of climate scientists on this blog. Many of the scientists are then sent threatening and bullying messages from his readers. On one occasion, Morano posted two email addresses for the highly respected (and sadly now deceased) climate scientist Stephen Schneider under a photo of Hitler. This incident occurred after Schneider's name appeared on a neo-Nazi website's 'death list' alongside other climate scientists with Jewish ancestry. Schneider said he had observed an 'immediate, noticeable rise' in threatening emails whenever climate scientists were attacked by prominent right-wing US commentators. 'What do I do?' he asked in exasperation. 'Learn to shoot a magnum? Wear a bulletproof jacket? I have now had extra alarms fitted at my home and my address is unlisted.' Like Hans Schellnhuber, the German scientist threatened with a noose in Australia, Schneider believed it was only a matter of time before a climate scientist was killed. 'I'm not going to let it worry me . . . but you know it's going to happen.'

It's not just climate scientists who are concerned about Morano's tactics. One of the *Guardian*'s editors, Leo Hickman, wrote: 'What possible reason could Morano have for prominently displaying these email addresses—as he does for many other stories that involve climate scientists he evidently despises—other than to encourage his readers who lap up his warped world vision to "get in touch"? I'll let you fill in the gaps.'

Morano has even made statements that could be interpreted as encouraging violence against climate scientists, saying 'they deserve to be publicly flogged'. On another occasion he told a TV news outlet, 'They deserve the public wrath they're getting. It's refreshing that they're finally getting a hostile reaction.'

Morano is involved in a raft of conservative causes. He is assistant treasurer of an organisation responsible for a bizarre 25-minute political attack ad that accuses President Obama of accepting campaign funds from the political movement Hamas, which is classified a terrorist group by the United States. The ad claims that Obama views America 'with an outsider's point of view, often with hostility'.

The journalist I'm having an email exchange with warns me of some of this. He strongly urges me to refuse to debate Morano. 'Tell them that you will not allow your reputation to be used to legitimise such a man or such tactics,' he writes. 'There is no "debate" possible with Morano. He just engages in non-stop lies and character assassination. TRUST ME ON THIS.'

I can't believe Nick has chosen Morano, given that the documentary we're involved in is called 'I Can Change Your Mind'. I've selected participants who have a fighting chance at doing just that—changing Nick's mind. They're highly respected in their field, have unquestionable integrity and many of them speak to Nick's own values. How could Nick possibly think that taking me to meet Marc Morano could change my mind? Even if Morano *accepted* climate science, I still wouldn't want his tactics on our side.

So I'm faced with a dilemma. Can I pull out of debating Morano? It sounds like I should, if only to make a point about my strong opposition to bullying in the climate 'debate'. Yet if I refuse to meet him this section of the documentary will become about how I was afraid to take on one of Nick's picks. So I decide on a compromise. I'll do the meeting but tell Nick at the start that because of Morano's reputation for dishonourable tactics and misrepresenting climate scientists, I won't enter into a point-to-point debate on scientific issues. I'll let him make his points and let the audience decide if he's a credible source or not.

I'm feeling nervous about this course of action. But I also feel it's the right thing to do. To enter into a genuine debate or dialogue with someone requires respect on both sides. Morano's actions in turning the climate change debate into a place of ugliness and hostility shows he has no respect for climate scientists or advocates.

Nick and I wait at a large stone memorial in front of the iconic Capitol building. Morano is late, but eventually we see him striding over. He's a large man in his early forties with bushy brown eyebrows, a sharp suit and very white teeth. Morano sees that Nick isn't wearing a tie so he takes his off. And then we begin.

'Anna, I was keen for us to come to Washington because I wanted you to meet one of the United States' leading global warming sceptics, a fellow called Marc Morano,' says Nick. '[He] has a website called *Climate Depot* and used to work for Senator James Inhofe, probably the Congress's leading global warming sceptic.'

My heart is beating fast. I take a deep breath and dive straight in. 'Nick, I wanted to say—before we start the substantive discussion— I'm quite surprised that you brought me to meet Marc,' I say. 'I know in Australia, people haven't heard of him, but in the US I feel like he's actually been quite discredited. He misrepresents scientists. He's been shown to just make things up. And most worryingly to me he publishes scientists' email address on his website under pictures of Hitler and encourages his readers to then email them, and then they get death threats.'

There. I've said it. That wasn't so hard. Nick is looking annoyed but it's not the end of the world. I continue. 'So I think that . . . I'm happy to listen to what Marc has to say today but I'm not going to engage in point-to-point debate. I feel like those kinds of tactics like encouraging violence against climate scientists have no place in this debate.'

Morano responds along the lines I'm expecting. 'What a cop out, Anna. You say you don't want to debate. Is that because you don't know if you have the science on your side?' he asks. 'I *proudly* post the publicly available emails of scientists who make absurd claims, who refuse to debate and who get nothing but fawning media coverage from all mainstream media, who get nothing but respect and civility from their students, who get nothing outside of their insular bubble.'

Morano is animated as he talks. His background in right-wing talk shows is obvious. 'What's so refreshing is, these scientists for the first time are hearing directly from the public via their website's publicly available email,' he says. 'I just help bring the public to these scientists who can finally hear the anger, the frustration, the intellectual arguments against them . . . You should be thanking me for publishing these email addresses.'

I turn to Nick. Maybe he hasn't understood why I'm objecting to the role Morano has played in taking the climate conversation to such an ugly place. 'I feel, Nick, that the tactics that Marc uses don't have a place in a legitimate debate. As you know I'm happy to debate any legitimate climate scientist but someone who consistently misrepresents scientists . . .'

'What's your evidence for that?' asks Morano aggressively. Nick joins in: 'Yeah, what do you mean misrepresents scientists?'

There are numerous examples here. 'I would think that the viewers of this program can do a quick Google search and find any number of misrepresentations,' I say. One such case happened when Morano was a staffer for Senator Inhofe. Morano issued a misleading press release about the work of a group of Swiss scientists led by Anja Eichler from Switzerland's Paul Scherrer Institut. Her research showed that the sun was a major factor in climate changes that happened millions of years ago, and stated clearly that solar activity was *not* the driving force for the climate change that has happened in the past 150 years. Morano's press release erroneously claimed that Eichler's study proved that half of global warming was due to the sun. 'Our conclusions were misinterpreted and we are a bit concerned about that,' Eichler stated after learning of the press release.

When I bring up the claim that he misrepresents scientists, Morano seems to think I'm talking about a report he compiled when he was working for Senator Inhofe. 'Any document that's 255 pages is going to have some addendums,' he says. I guess he must have been criticised for misrepresenting scientists in that report, too. But Nick doesn't seem to mind. 'Whatever,' he says. 'It doesn't begin to compare to the fraud in Al Gore's movie *An Inconvenient Truth*. That was the greatest fraud ever perpetrated on the global public so really, let's get real.' Hmm, I think. Perhaps the sub-prime mortgage crisis and systematic corruption of the US financial services industry might rate higher than Al Gore's movie on the 'greatest fraud' list but anyway. . .

Morano goes on to claim that the sea level hasn't risen but rather dropped. 'Contrary to predictions of it accelerating [sea level rise] has actually dropped, an historic drop according to the European Space Agency and NASA,' he says. He also claims that Antarctica is gaining ice and that 'ocean heat content [is] dropping or stable since 2003'. I *know* that all these points are blatantly inaccurate, but I've decided not to enter a point-to-point debate. 'Anna, you should not be affiliated with a movement that proved itself [based on] subprime science, subprime economics and subprime politics,' Morano says. 'You're going to be entering the dust bin of history. History's passed you by. I feel bad,' he says.

'Here's the whole history of the environmental movement,' continues Morano. 'Almost every eco scare of the day had the same solution: more government, international global governance, more bureaucratic bean counting, more individual control over individuals' lives.' I'm incredulous at this melodramatic allegation, but in a way also not surprised. Morano is revealing the same motivation that Naomi Oreskes' research ascribed to the original gang of physicists who founded the George C. Marshall Institute. It's the fear that concern over climate change is really some kind of communist plot to curtail individual freedom.

'It's the same impulse,' says Morano. 'Whether the eco scare is over-population, deforestation, it's always more control over our lives.' Morano *definitely* has issues around government control. 'They're going to be able to control everything,' he says. 'And this is why Al Gore brags about the climate treaty, the global governance.' He stares at me angrily. 'You are the face of one of the greatest threats of our liberty,' he tells me. 'And that is intellectual, international bean counters trying to control average people's lives.'

'You've been had,' he says. 'You need to go back and re-examine your conscious, Anna.'

'My conscious?' I reply. 'I'm pretty conscious and I have looked at the evidence.'

'Conscience,' clarifies Morano. 'Your intellectual conscience.'

Morano tries hard to get me to debate scientific points with him. 'Debate the issue,' he says. 'Tell us what you know, instead you just want to, like, "Oh, I'm out, I'm out, I don't have to say anything."

That's weak, it's cowardly . . . Basically she's saying I'm a liar so she won't talk to me.'

Nick says: 'Look, it's a real problem in this debate that those of us who are, you know, not prepared yet to be convinced that man is causing dangerous global warming just get attacked personally, all the time. I don't understand that.'

At this point I have to jump in. How can Nick be so hypocritical? 'Nick, how can you say that, yet have brought me to the worst of the worst Republican attack dogs who constantly slurs people's reputations?' I ask. 'I would have thought that you of all people would pick a different kind of spokesman to take me to in Washington.'

'You just don't like what he says, you just don't like what he says,' Nick interrupts, defending his choice. 'I have enormous respect for Marc. I think he runs one of the most interesting websites there is on this issue. He's done enormous work in Congress on exposing a lot of exaggeration,' he says. 'I applaud Marc for what he's doing in exposing this exaggeration because people aren't getting the facts. They're not getting the facts that, you know, the planet hasn't actually warmed for thirteen years and yet CO_2 is still going up.'

I groan. Nick is still on the 'no warming since 1998' bandwagon. Yes, 1998 was an unusually hot year due to a strong El Niño, but 2005 and 2010 were slightly hotter and the past decade is easily the hottest on record.

We finish up the conversation by Marc telling me that climate change is over and the next bandwagon for environmentalists will be species and biodiversity. 'When you switch over to species and I come back in three years, and you're talking about species being the greatest threat and you've forgotten about global warming, learn your lessons,' he says. 'And don't make it the apocalyptic one.'

I highly doubt that in three years I will have stopped caring about climate change, but Morano seems convinced this will happen. I tell him it really won't. I've cared about this issue since high school.

This sets Morano off on another rant. 'Schoolchildren are all the biggest believers, that's the real tragedy of this,' he says. 'Children under eighteen. Polling has consistently shown they're the biggest believers in this.'

Well, maybe that's because it's their futures at stake, I think. They're paying more attention.

But Nick has an alternative explanation. 'Because they're getting all this stuff at school all right,' he says. 'Terrific, mate. It was terrific meeting you.'

After the meeting Nick and I wait for a taxi in silence. I think about what I've done. It might work well on TV or it might work badly. The way I see it, I took a principled step and confronted a bully. In light of all the climate scientists whose lives have been affected by his attacks, I feel it was my responsibility to say something. I called him out on behaviour that should be unacceptable in a civilised debate. And even if it doesn't play well with the television audience, I know I've made the right decision for myself. I'll probably get attacked for being so blunt about my feelings towards Morano, but I wouldn't forgive myself if I helped legitimise someone who has instilled so much hatred into the climate debate.

Chapter Thirteen

IN THE NAVY

I'M FEELING DISHEARTENED AFTER MEETING MARC MORANO. I REALISE I'd thought Nick was a different kind of climate sceptic, someone who'd been open to a more civilised argument. I certainly hadn't expected he'd be interested in siding with a character like Morano, who bullies climate scientists and makes ridiculous claims about sea levels falling when they're actually rising.

I was naive, I guess. Nick seems happy to align himself with anyone who will argue that human-induced climate change isn't real. Nick confirmed this at the end of the meeting when he said, 'I have enormous respect for Marc.'

I also feel let down, because it seems that I'm the only one playing by the rules as outlined to me at the start of the project. I chose my spokespeople because they could reach out to Nick and have a chance of *actually changing his mind*. They're moderate and polite. If anything, they understate the seriousness of climate change in order not to scare off conservatives like Nick.

I could have taken a different approach and chosen NASA's chief climatologist, James Hansen. He's probably the world's most eminent climate scientist and in recent years has also evolved into an activist. Like many scientists, he's now willing to be outspoken about the implications of the science for public policy. Hansen would have got stuck into Nick's claims in a way that may have won the debate with the TV audience, but embarrassed Nick and closed his mind even further.

There are any number of eloquent climate activists who speak regularly and powerfully about the causes and impacts of climate change. The writer and co-founder of 350.org Bill McKibben, the head of Greenpeace International Kumi Naidoo, or the British author and intellectual George Monbiot all would have demolished Nick's arguments head-on. Sparks would have flown. Bill, Kumi and George know the blocking role Nick has played in stopping climate action. And they wouldn't have been polite about it.

Instead I'd opted for a 'softly, softly' approach with my spokespeople. I hadn't chosen activists or media performers. I thought they would have closed Nick's mind even further by being too aggressive towards him. Instead I'd chosen diplomatic spokespeople who are mostly not part of the climate change campaigning world.

On the train from Boston to Washington DC, Nick and I had debriefed the journey so far. We'd discussed the merits of our respective spokespeople.

Nick had claimed that Dick Lindzen was the best climatologist in the United States. I'd called this proposition 'extreme', given that Lindzen's work on feedbacks has been criticised severely by his peers for decades.

'Tell me who's more respected in the United States?' asked Nick.

'Well, people who have won the President's Medal for Science for example, like Susan Solomon from NOAA,' I said. 'Or James Hansen from NASA.' Nick exploded in anger.

'Hansen? Hansen's a joke. I've had enough of this conversation,' he said. 'I can't stand Hansen. I'm out of here. Hanson's a fraud.' Nick shifted uncomfortably in his seat and I expected him to storm off, but we were on a fast-moving train. There was nowhere to go.

'What about his work do you object to?' I asked politely, taken aback at his over-the-top reaction to the mere mention of NASA's chief climatologist.

'He's a complete fraud and I wouldn't go and see him either,' said Nick.

Perhaps I *should* have selected James Hansen.

I compare my spokespeople to Nick's choices. By now I know the full list of the people he's taking me to meet. Five of Nick's six spokespeople are professional climate deniers. They're seasoned media performers who spend most, if not all, of their day jobs arguing against climate science in the public sphere. Nick has been strategic in his selections, that's for sure. But it doesn't seem he has chosen them based on their ability to change my mind.

The person I'm taking Nick to meet now is as moderate and conservative as they come. There's no way Nick can write him off as an 'alarmist'. Rear Admiral David Titley is the chief oceanographer of the US Navy. He mostly works out of the Pentagon. On any other matter, Nick would take his views extremely seriously.

The documentary's executive producer, Simon Nasht, had been the one to suggest Admiral Titley. Simon knew that Nick had two sons in the military and was aware of Nick's respect for the armed forces. And the Admiral has a compelling story. He is a former climate sceptic who now accepts the science. He heads the Navy's task force on climate change. He closely monitors ocean-related elements of climate change such as sea-level rise and the opening up of the Arctic passage due to rapidly melting ice.

Still, the Admiral hasn't done much media work. Unlike Morano, whose television appearances are all over YouTube, the Admiral hasn't done a single television interview that I can find. A military-affiliated contact of mine in the Unites States, Kevin, warned me that serving officers in the US military tended to be extremely cautious when talking publicly about climate change. For this reason he recommended using a retired officer as a spokesperson rather than one currently serving. Despite Kevin's advice I ended up going with Admiral Titley as I didn't have time to organise another military spokesperson. The production team had already put the request into his office and I crossed my fingers that he'd come across well to Nick and on camera.

•

An hour after the Morano incident, it's still tense between me and Nick. The security is so tight at the Naval Observatory that they won't let our cab driver through the front gates. So we get out and climb into the van Max has hired, instead. When we are finally cleared to go through the gates, the Naval Observatory comes into view. It's an impressive building with a large white dome at the top. The building is surrounded by an expanse of lush green lawns complete with a frolicking herd of deer. The vice-president's residence is just around the corner. As we walk inside the observatory, we see a tall staircase that leads to the Admiral's office.

Nick and I wait downstairs with Kate for what seems an eternity. Max, Pete and Leo set up the cameras and lights upstairs in the Admiral's office. I sit down on a wooden bench and focus on the conversation we're about to have. Nick paces the hallway, looking at old Navy photos. Eventually, Max appears at the top of the stairs and beckons us up.

Admiral Titley's office is enormous, with a huge wooden desk in the centre flanked by two official flags. Bookshelves line one wall and there's a collection of Navy paraphernalia arranged around the room. It feels like we're in a scene from *The West Wing*. This is his official office. After our conversation, he's heading back there for a meeting with some high-ranking generals.

Admiral Titley is tall and fit, with a light tan and greying hair. He's wearing what the Navy call 'service dress' uniform: a double-breasted black jacket, white shirt and black tie. 'Call me Dave', he says. But the environment is so formal and his uniform dripping with so many badges and medals that it just doesn't feel right to call him anything except Admiral.

The Admiral tells us he's always had a deep interest in weather and climate. When his family moved house when he was five years old, he chose his new bedroom simply because it had a thermometer outside.

He went on to study meteorology at Penn State University, joined the Navy, and spent the next three decades doing what he calls 'operational oceanography and weather'. The Navy is understandably interested in long-range weather forecasting. They need to know the conditions under which they'll be carrying out their missions. Will there be clear skies or rough seas a few months from now? It makes a big difference to the captain of a Navy ship.

The Admiral says that at first, he wasn't really interested in all the talk about climate change. 'I have to admit,' he says, 'when I first heard about this and started to think about it, I thought back to, well jeeze, those weather forecasts after day three, day four they just they weren't very good.' He tells Nick and me how hard it is to forecast weather.

'So I was listening but I can't say I was convinced,' he says. 'You could probably call me a sceptic. Not a sceptic as in "well this just can't be true" but as in, "I'm really not sure."'

This changed when the chief of naval operations asked Admiral Titley to take on the job of chief oceanographer. Part of the role involved heading a task force exploring the impact of long-term climate changes in the Arctic. 'So I really started taking a look at all the data,' he says.

Admiral Titley's approach to determining whether or not 'there was anything there' when it came to human-induced climate change was influenced by his former role as a navigator on Navy ships. 'One of the things that you learn when you're a navigator is you never trust any one single data source,' he says. 'Because that's how you can end up in places you're not supposed to be—like the rocks.' I glance at Nick, thinking of his insistent reliance on Lindzen's lonely stance in the scientific community.

'So you learn to take what's called dead reckonings, and celestial, and radar—and you will take all of these different data sources and you come up with a pretty good mental picture of where you are and where you're going,' he explains. 'I have used that model for trying to evaluate climate change.'

'We did a pretty exhaustive tour of many places,' he says, 'just looking at the data. 'And what the data was showing me was the ice was melting out of the Arctic at a rate we hadn't seen. The NASA temperature records, the British temperature records, some other temperature records were all either at or near a record high. The oceans were warming . . . And I had some people much, much smarter than me try to help me think through this.'

Admiral Titley's major concern about climate science was that if weather forecasts were so unreliable for one week in advance, how could climate models project temperatures decades into the future? But he tells us that one of his mentors emphasised the difference between weather and climatology. 'In climatology it doesn't really matter so much what the weather today or yesterday was. What matters is what's the relative difference between how much energy from the sun is coming in, and how much energy from the atmosphere is going out,' he says. 'And that, as we all know, has changed over the millions of years.'

'Our space agency, NASA, has put up a series of satellites that amongst other things actually can measure how much heat is leaving [the atmosphere],' he says. 'We see—right where the wavelengths for some of the greenhouse gases, like methane and carbon dioxide—we actually see *less* heat leaving the atmosphere now than we had before. So really the science is saying you have the same amount [of heat] going in and you have less going out at those wavelengths.' It was this data that led the previously sceptical Admiral to the inescapable conclusion. 'So over time the earth's system . . . has to be getting warmer.' Something had to be trapping the heat.

To confirm this, Admiral Titley compared the satellite evidence with what was happening in the real world. He asked, 'OK, that's what the theory says, so what does the data say? And that's where I get back to looking at all these different things: the ice, the thinning of the ice, the warming of the water, the warming of the atmosphere, [changes in] plants and animals.'

'So for planning purposes for the US Navy, we accept the scientific consensus that the climate is changing and that there's probably a man-made component,' he says. I cringe internally at the word 'probably'. I know he has to say it. The IPCC's last major report, in 2007, says ninety per cent probability rather than one hundred per cent, and he has to be very careful that his words reflect this, but I know that Nick will seize on it later. To limit the damage, I put to him Nick's argument that the climate change we're seeing is just natural variation, rather than human-caused.

'If this is just natural variation,' the Admiral says, 'what I would ask the . . . folks who propose that is . . . at what point does the natural variation stop? Does the temperature . . . come down? And at least the data sets that I've seen haven't really shown me that we're at that point.' He's saying that if climate change is just natural variation, decadal temperatures would be actually varying (i.e. going up *and* down). Instead, they're just going up and up with no foreseeable end point in sight.

Nick doesn't argue with the Admiral. I wish he would—I'd love to see the Admiral, who has a PhD in meteorology, correct Nick on the science. Instead, Nick asks a bunch of non-science questions, like why all this is important to the Navy. 'So you're seeing the Arctic as a key strategic area for future potential conflict?' asks Nick. In answer to this, the Admiral offers to take us down to the Naval Observatory's library to talk in more detail about the changes happening in the Arctic.

The library is incredible. It's a circular space surrounding a small, elegant indoor fountain. The floor is duck-egg blue. An enormous chandelier hangs above us, illuminating two levels of bookshelves filled with leather-bound books of all colours and sizes. Light streams in from decorated windows close to the ceiling, casting soft shadows on the globes, statues, maps, artwork and antique wooden desks on display. Many of the books date back to the 1400s, 1500s and 1600s. The three of us gather around a globe from the 1840s.

The Admiral explains what's happening in the Arctic. Normally, the Bering Strait between Alaska and Russia is covered in ice. 'In the summer, especially at the last few years, we've seen open water for hundreds of miles,' he says. The Navy is keeping track of the forty per cent drop in ice since the 1970s as well as the very significant decrease in the ice's thickness.

The Admiral tells us a story about when he was in the Arctic two summers ago. 'I was on the United States coastguard cutter *Healy*, which is an ice breaker. We came out of Barrow, which is our northernmost town in Alaska. We steamed for about a day and a half due north, and we found one tiny little ice floe. And that was it,' he says. Normally, you'd expect to find huge swathes of ice. 'We had one of the tribal elders on board and I asked him was there anything in their oral history about this. He said no. The water temperature was plus four, plus five. He said they'd never seen temperatures like that.'

The Admiral tells us about another time he was in Barrow recently. The temperature was 22 degrees Celsius. This is a place where it's normally well below freezing. 'I had brought all this winter gear,' the Admiral tells us. 'I was walking around Barrow in my shirt sleeves.'

'They must have loved that,' says Nick. 'The locals must have loved that.' Yet again, Nick manages to trivialise climate change with a glib line about how great the impacts of climate change are. But the Admiral doesn't laugh.

'Well, they wouldn't know what to do,' Admiral Titley says. 'The issue for much of the structures up there is it's all built on this permafrost.' Permafrost is the perennially frozen ground that covers half of the Northern Hemisphere. An article in the *New York Times* describes it as a 'storehouse of frozen carbon, locked up in the form of leaves, roots and other organic matter trapped in icy soil'. Much of Alaska's infrastructure and communities are built on permafrost.

This permafrost is now melting for the first time in over 120,000 years. For Alaskans, this means gaping cracks opening up in the earth. Buildings are being abandoned because the solid ground they were once built on is collapsing like rotting floorboards.

'So now the permafrost is melting,' says the Admiral. 'And you get these buildings that look like they're in an earthquake because they're all leaning. Now how do you deal with this?' he asks with a frown. 'Big, big changes for the communities up there.'

Nick doesn't respond.

In fact, the area of Arctic sea ice has been declining faster than climate models predicted. At the end of 2011, the end-of-summer sea ice was around forty per cent less than that in the late 1970s, when NASA's satellite measurements began (Figure 13.1).

James Hansen predicts that 'continued growth of atmospheric carbon dioxide surely will result in an ice-free end of summer Arctic within several decades, with detrimental effects on wildlife and indigenous people'.

While Rear Admiral Titley is thinking about what this means for the Navy in terms of short-term commercial shipping, NASA is thinking about what it means for the climate long term. Once the Arctic goes, the Greenland ice sheet will likely melt, and we've passed a tipping point. If the permafrost melts too, that's a double whammy: another tipping point.

Figure 13.1 Sea ice extent

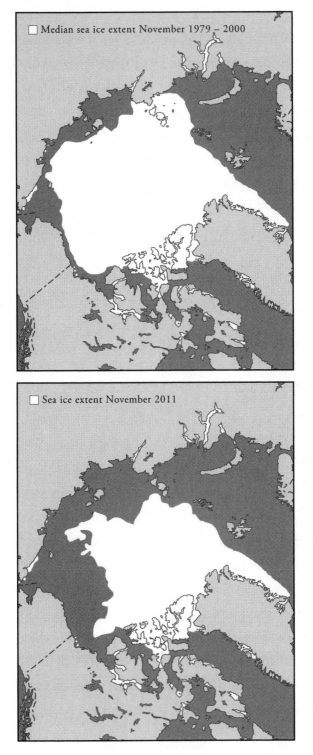

Source: Adapted from National Snow and Ice Data Center, http://nsidc.org/arcticseaicenews/2011

When the permafrost melts it releases the greenhouse gases it absorbed then froze during the last ice age. When the ground was frozen, these gases were locked up in the equivalent of a safety deposit box.

The melting unlocks the gases. 'It's like broccoli in your freezer,' said Kevin Schaefer, a scientist at the US National Snow and Ice Data Center in an interview with the *New York Times*. 'As long as the broccoli stays in the freezer, it's going to be O.K. But once you take it out of the freezer and put it in the fridge, it will thaw out and eventually decay.' And for the permafrost, once the decay begins it won't be able to stop. A paper from the National Snow and Ice Data Center concludes that the thawing of the permafrost is definitely a feedback mechanism. 'The thaw and release of carbon [from the permafrost] . . . will change the Arctic from a carbon sink to a source after the mid-2020s. The thaw and decay of permafrost carbon is irreversible,' it states.

If this happens, it's game over for the climate. The permafrost is often referred to as a 'carbon time bomb'. It holds within it twice as much carbon as the entire atmosphere. The scientific journal *Nature* reported that 'the latest estimate is that some 18.8 million square kilometres of northern soils hold about 1,700 billion tonnes of organic carbon . . . That is about four times more than all the carbon emitted by human activity in modern times and twice as much as is present in the atmosphere now.'

The climate implications of a world with disintegrating permafrost and no Arctic summer ice are devastating. Nick and the Admiral are talking about what it means for commercial shipping routes. Nick seems excited that global commerce can save time, and therefore money, on shorter shipping routes through the Arctic. 'So there are commercial vessels currently doing that summer voyage?' Nick asks enthusiastically. But I'm thinking about the implications for feedback mechanisms when the ice is gone. So is the Admiral.

He tells us that when air temperatures are warmer, white ice melts, revealing dark ocean that triggers a feedback mechanism. 'As you can imagine, white ice reflects the sunlight, whereas the very dark water absorbs that sunlight,' he says. 'So the more water you have the more [heat] energy you're able to put into the water. That warms up more of the water therefore you melt more ice, and so on. It builds upon itself really quite rapidly.'

This is why NASA's James Hansen had said, 'It is difficult to imagine how the Greenland ice sheet could survive if Arctic sea ice is lost entirely in the warm season.' And if Greenland melts, we're looking at global sea level rise of six to seven metres.

Even without the Greenland ice melting, the Admiral understands that sea level rise will be a major problem. 'I think by the year 2100 compared to 2000 we're probably in for at least a metre of global sea level rise,' he tells Nick and me. 'I think if I'm wrong, unfortunately I'm probably wrong on the low side.'

None of these scenarios seems to be playing out in Nick's head. On the contrary, Nick seems to be having a great time in the Naval Observatory Library. He's cheerful and talkative. He and the Admiral are getting along well. With Nick's knowledge of the military, they chat about naval operations, shipping routes and United States–Russian cooperation over the future of the Bering Strait. It's all getting a bit blokey.

I'd been hoping the Admiral could talk about the global security implications of climate change. One of the many books I have in my suitcase is the book *Climate Wars*. The author, Gwynne Dyer, writes about how an increase in average global temperature of two degrees could 'heat global politics to boiling point and trigger massive conflicts over scarce food and water'. He states that once these resource wars begin, 'all hope of international cooperation to curb emissions and stop the warming goes out the window'.

In 2007, the British foreign secretary at the time, Margaret Beckett, admitted that the consequences of climate change 'reach to the very heart of the security agenda'. I ask the Admiral about these implications for global security.

'The types of things that we're looking for is, does climate change act as one of the forcing functions that may take an unstable world and make it even less stable?' he says.

'And I think how much that happens is very much up for debate, but clearly the small Pacific Island countries like Tuvalu, places like that, as sea level comes up, I mean they're going to have some tough choices,' he says. 'Do you in fact need to move to other locations?'

The Admiral shares some questions that fall into the category of 'what keeps me up at night'. 'Is there a minimum temperature threshold below which [you get] decreasing rice crop yield? Do the monsoons really and truly move at some point? Is there something that puts the climate into almost another . . . mode, if you will?'

He's talking about tipping points. 'Is there anything that's going on that would cause us to get into that much more unstable [climate] mode?' he asks. 'I'm not saying it's going to happen . . . I don't know.'

The Admiral is concerned about the food security implications of climate change. 'If you start moving where you can do agriculture, not by 20 kilometres but let's say you start moving that by hundreds of kilometres . . . it would be a very different world,' he says. 'How we would adapt to that kind of world I think is an open question.'

It makes sense that a military officer would think about this. As Gwynne Dyer wrote in *Climate Wars*: 'Eating regularly is a non-negotiable activity, and countries that cannot feed their people are unlikely to be 'reasonable' about it. Not all of them will be in what we used to call the 'Third World'— the developing countries of Asia, Africa and Latin America.'

His point is that 'the most important impact of climate change on human civilisation will be an acute and permanent crisis of food supply'. This is exactly why I began this journey by taking Nick to my uncle's farm.

•

Before we finish I want to quickly run by the Admiral the claim that Nick and I heard from Marc Morano a few hours ago. 'We saw someone this morning who said that sea level is dropping rather than increasing,' I say. The Admiral looks incredulous. 'I don't know what the data is they're referring to,' he replies. 'All the data I've seen says that sea level globally is coming up. It's coming up about 3 to 3.5 millimetres per year; it's about fifty per cent higher than it was.'

We finish up by again talking about risk. 'I read a great quote from General Gordon Sullivan,' I say, referring to the former general in chief of the US Army. 'And he said when you wait for a hundred per cent certainty on the battlefield, people die. And we don't want that to happen with climate change.'

'Yes, I mean it's too late,' replies the Admiral. 'Nobody who wins a battle probably had a hundred per cent certainty. I mean it just . . . you know . . . life just does not work that way.'

The meeting ends and we go outside. Max films Nick in a debrief while I sit on the grass and wait. A few deer approach the ground in front of me and I think about Harry Potter's Patronus charm. It's a spell that evokes a positive energy force, tangible in the form of an animal, which dispels fear and instils confidence. Harry Potter's Patronus is a stag, which is why I'm reminded of it. What would mine be if I had one? At the moment, it might be a polar bear to crash-tackle Nick and remind him of the consequences of the melting Arctic.

I have no doubt Nick will be gloating in his debrief about Admiral Titley's use of the word 'probably' in his comment that climate change is 'probably caused by humans'. The slightest doubt fuels climate scepticism. With that in mind, I'll put this day down to a win for Nick.

When we get back to the hotel I run through the events of the day in my head. Rear Admiral Titley was knowledgeable and authoritative, a sharp contrast to Marc Morano. Despite this, I'm disappointed at how the Navy meeting went. I know in my heart that the Admiral didn't make the kind of short, snappy, strong comments that are persuasive to a TV audience. To take on someone like Nick and perform well—not just in front of him, but for the audience who'll ultimately view it—you need to be willing to make strong statements that aren't peppered with qualifications.

This is why so many climate scientists are unwilling to go into public forums to 'debate' the science. Their academic need to qualify every statement means that they're out of their depth in the era of the snappy sound bite. I learned at a conference a few years ago that a sound bite is supposed to have twenty-seven words, three thoughts, and be eight to nine seconds long. Maybe I should have mentioned that to the Admiral beforehand.

NASA's James Hansen has spoken about the problem of scientific reticence before. 'What can we expect the public to think when they compare a scientist who includes appropriate caveats with a contrarian who gives conclusions without hesitation?' he said. 'It can seem like a debate between theorists, and often contrarians are more media savvy.'

Then I get annoyed that I even *have* to judge my spokespeople on how they perform on television. I want to have a real conversation, not a polished media performance. But the nature of the program and presence of the cameras are getting in the way of that. Ultimately the end product is a TV show, and both Nick and I are very aware of that. Nick's media training as a politician and my own media experience probably aren't helping.

Weeks ago, at Perth airport, I'd suggested to Nick that if we still hadn't agreed on the science halfway through we should lock ourselves in a room without the cameras. We could work through each part of the science with as many textbooks, scientific papers and phone numbers of respected climate scientists as we needed. He'd laughed, even though I told him I was serious.

I briefly consider bringing up this suggestion for an all-nighter science session again. But by now I've realised that Nick's mind probably cannot be changed. There's an invisible line in the sand for him that stops just before the science and starts with his views on the free market and corporate regulation.

Chapter Fourteen

ALL IN THE MIND

WE ARRIVE IN NEW YORK. IN THE TAXI FROM THE TRAIN STATION, OUR driver tells Nick he loves New York in summer because the women wear short skirts. Nick chuckles. To Nick, the taxi driver's comment is further confirmation of the argument he likes to make about hotter climates being better places to live. I look down at my knee-length green dress and decide to change into jeans for our afternoon meeting.

I drop my bags in my room. The window doesn't open and there's barely enough space for my bags. I don't mind though, because we're in New York City! I'm getting ready to go for a walk to explore when I get a phone call from Max. He reminds me I'm supposed to come to room 1104 in half an hour for our next meeting. This conversation is supposed to be with another 'neutral person' planned by the production team. Neither Nick nor I are supposed to know who we'll be meeting with.

I'd figured out the subject of the meeting a few days ago after a change in our itinerary. A planned stop in New Haven on the way from Boston to Washington DC had been cancelled. Max hinted to me and Nick that we might do the meeting later via Skype. New Haven isn't a big place. The entire economy is centred on Yale

University. I went through my mental catalogue of climate researchers and the answer popped up. We must be meeting someone from the Yale Project on Climate Change Communication. The project is world-renowned for producing excellent research into public opinion on climate change. I'd seen a paper they'd updated recently called 'Climate Change in the American Mind'. It would make sense, I thought, for the producers to have selected one of the research team to talk to us about their findings.

At the designated time, I knock on the door of the hotel room Max has booked for the video conference. Nick is already inside. The room is set up with lights, camera and sound equipment. There's a desk in the middle with a laptop, iPad and two chairs. I sit down.

Our first task is to complete a quiz on the iPad. The quiz is based on the one used by Yale in their research report. It will reveal where Nick and I fall on the spectrum of American attitudes about climate change.

Nick has trouble using the iPad so I help enter his answers. He selects 'I don't know' to a lot of questions. He really should be saying 'no', I think—for example to the question, 'Do you think climate change is caused mostly by human activities?' But it's part of his 'remain to be convinced' message. To delay action on climate change, Nick doesn't have to oppose the evidence outright. He just has to say he isn't yet satisfied there's enough of it to decide to do anything.

After we both finish the quiz, the results include notifications of where we sit on a spectrum based on our answers. I'm in the 'alarmed' camp, made up of people who 'are convinced that climate change is happening, that humans are the main cause, and that it is a very serious threat'. The survey tells us that 'the alarmed' tend to be moderate-to-liberal Democrats who are active in their communities. Demographically they come from every segment of society, but are more likely than average to be women, middle-aged, college-educated and upper income.

'Doctor's wives perhaps,' says Nick. 'It's an Australian analogy.'

'Well, if they're college educated they're probably doctors . . . that's a bit sexist Nick,' I say. To assume that an educated, high-income woman must be a doctor's *wife* rather than an actual doctor irritates me.

'Oh gee, lighten up,' says Nick.

Nick falls into the 'doubtful' category, although personally I think he belongs with the 'dismissives'. The survey reports that people in the 'doubtful' group are evenly divided among those who believe that global warming is happening, those who don't, and those who haven't decided yet. For the people in this category who *do* accept global warming is happening, they mostly say it is due to natural changes in the environment. 'Most are not very worried about it and say it is not an important issue to them personally,' says the report. Demographically, the 'doubtful' are 'more likely than average to be older, white, male, higher educated and higher income. They tend to be conservative Republicans and religious.'

With the quiz completed, it's time to meet the researcher who conducted the study with a nationwide sample of Americans. Dr Anthony Leiserowitz is director of the Yale Project on Climate Change Communication and a research scientist at Yale's School of Forestry and Environmental Studies. He has served as a consultant to the School of Government at Harvard University, the United Nations Development Program, the Gallup World Poll and the World Economic Forum.

Tony appears on the screen with headphones and a smile. 'Great to meet you both,' he says cheerfully. He explains the context and findings of his research. The 'six Americas' he found to exist on climate change, he says, are divided into the alarmed, the concerned, the cautious, the disengaged, the doubtful and the dismissive.

Figure 14.1 Global Warming's Six Americas

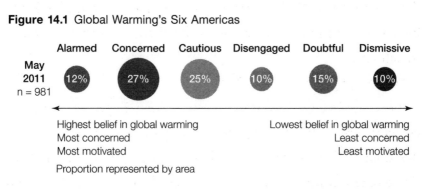

Source: Yale/George Mason University

We start by talking about the doubtful and dismissive categories. When we met Professor Naomi Oreskes earlier in the trip she told us about how the origins of climate denialism are deeply rooted in a fear of government regulation. This anti-government fear was born from the suspicions and apprehensions of a group of conservative Cold War scientists who went on to form the George C Marshall Institute.

At the time I had wondered if fear of government regulation was still a primary motivator for people who reject climate science today. Now I learn from Tony's research that this is indeed the case. Tony tells us that the six groups are primarily defined by 'their underlying values'. He tells us that the groups most concerned about climate change tend to have very strong egalitarian values. 'They're very concerned about social justice, environmental concerns around the world, the impact on other people and places,' he says. 'Fairness is really a core value for them.'

The doubtful and the dismissive, however, are what Tony calls 'strong individualists, almost libertarians. They often have a very anti-government view,' he says. 'They believe that the government should get out of our lives and should have as little interference in society as possible.' Tony makes the point that 'much of what we see going on, for instance, in terms of hostility to the climate change issue,' is not about science. He says that many people in the dismissive group 'talk about climate science but for them it's not really about the science. They're deeply afraid that climate change is really a pretext for more government interference in society.'

Tony explains why this is the case. 'So in other words what they're afraid of is the potential *solutions* to climate change, much more than climate change itself,' he says. 'Whereas of course the alarmed and the concerned are totally focused on the issue of climate change, the scientific idea and the fact of climate change. So you've got these two very different conversations going on at the same time.'

I wish Nick would respond directly to this. To me it's so clear by now that this is the reason he rejects climate science. But instead Nick changes the subject. 'Did you profile them by voting intention?' he asks, moving away from the issue of values.

We move on to talk about the quiz itself. Nick has some problems with questions that he felt should have a 'don't know' option. He particularly didn't like the question that asked how easy it would be to change your mind on climate change.

'You had the question, could you easily, I think it was, change your mind or something. And I said—well hang on a minute, I don't easily change my mind about a subject like that. So to me the question was difficult to answer,' he says. 'I can easily change my shoes but I can't easily change my mind about a big public policy issue like this.'

Nick's objection to the question seems to answer it. He simply isn't open to changing his mind about climate science. As both Professor Naomi Oreskes and now Professor Tony Leiserowitz have pointed out, he has good reason for this position. This debate isn't really about science for him, but rather his core values.

He believes government should be small and hands off, and markets left to their own devices. Climate change demands structural changes to our economy and energy systems, including regulating energy corporations. Therefore, for Nick to accept it as real and human-caused poses a major challenge to his world view.

I can understand why this is difficult for Nick. I imagine what it would be like if some new piece of evidence came to light that said the only solution to climate change was to slash the world's foreign aid budget in order to spend it on renewable energy. I care deeply about ending poverty so I would be horrified. The first thing I'd do is look for some other solution.

But what if I wasn't so aware of climate science? To protect the foreign aid budget, maybe I would reject the evidence for the *problem* the cuts were supposed to solve. That way, I could argue against cutting the aid budget on the ground that there was no problem that needed to be fixed in the first place. When I put myself in his shoes like this I almost feel sorry for Nick. He has a lot at stake in maintaining there's nothing wrong. For me to ask him to accept climate science is to ask him to give up something core to his identity and values.

Tony, Nick and I talk over Skype for around an hour. Tony has some interesting insights. When we'd met Marc Morano a few days ago, Nick had been delighted to point out that climate change has fallen down the list of priorities for the American public. He asks Tony about this. Tony agrees with Nick's analysis and offers an explanation. Media coverage of climate change in the United States has dropped by about sixty-five per cent since 2007. Nightly network news coverage of climate change dropped even further, by eighty per cent.

'Now that's crucial,' he says. 'Because we can look out the window and there's CO_2 pouring out of the tailpipes and the smoke stacks and out of the buildings. But we can't see it. It's invisible. And likewise the impacts are invisible to most of us who don't know where to look.' Tony points out that the public 'don't read peer-reviewed literature. They don't know scientists . . . So when the media doesn't report this issue it's literally out of sight and out of mind.'

We briefly talk about confirmation bias. This is a psychological phenomenon that Nick and I, like all humans, are both susceptible to. 'We all have this tendency to look for information that validates what we already believe and to ignore or to discount information that contradicts what we currently believe,' explains Tony. 'And even most subversively, we take information that *does* contradict what we believe and distort it and twist it and say no, that really does [prove my position].'

So is there any hope for conservatives like Nick accepting the facts of climate science? Some have values preventing them from doing so, and confirmation bias then kicks in to validate their incorrect position. I don't ask this question out loud but Tony seems to sense it and answers it anyway.

'That exact human characteristic [of confirmation bias] is exactly what science as an institution evolved to try to address,' he says. 'Scientists also come up with their own stories about how the world works.'

But then he tells us that in science, a story is not enough. 'You have to provide evidence and you have to tell a compelling logical argument and most importantly it has to be *replicated by other scientists*,' he says. 'And if we can find convergent evidence—evidence from all sorts of other fields that all seem to converge on the exact same conclusion that's even stronger. That's what the institution of science at its heart is all about.'

Like most people we've met with so far, Tony brings up the issue of risk. He asks about the level of certainty required before governments act to cut carbon pollution. 'There's always going to be an element of uncertainty,' he says. 'We'll never have complete knowledge about climate change. But in fact we don't have complete knowledge about *any* of the important decisions we make.'

Tony uses the example of economic policy. This is an example that's close to Nick's heart as a former finance minister. 'Australia is making huge decisions about . . . its industry and its regulations and its laws [and] what effect they're going to have in the future. But you don't know. Nobody knows,' says Tony. 'That's why we buy insurance. It's why we have all these institutions that help us protect ourselves against worst-case scenarios. In that sense I think that [risk management] is a much more productive way to think about climate change.'

I turn to Nick. 'So how risk adverse are you in other areas of your life?' I ask.

'I don't think that's particularly relevant to the public policy issue,' replies Nick.

'Well it might be,' I say. Maybe Nick is just generally someone willing to take a lot of risks on everything.

'Well, no,' says Nick. 'No, I don't think . . . whether I'm prepared to walk in the streets, you know, of New York at 2 am—my risk assessment of that particular course of activity has nothing to do, really, with my assessment of the risks or otherwise associated with the continuing production of CO_2 through human activity. I don't see the comparison whatsoever. Sorry.'

Nick and I do manage to get some agreement on climate change solutions that are worth doing regardless of the cause or scientific foundation of climate change. We both support energy efficiency and moving to clean energy. We just disagree on the timeframe for the transition from fossil fuels to renewable energy sources.

I argue that Australia can and must switch to one hundred per cent renewables within a decade. The climate can't afford for the world to wait. Research from the Melbourne-based think tank Beyond Zero Emissions proves that not only is this necessary to reduce our emissions to safe levels, it's also feasible using current technology with a mix of concentrated solar thermal, wind turbine, photovoltaic solar, hydroelectric generators and biomass energy.

Nick is more relaxed about the timeframe of the transition to renewables. He thinks that it will be a long time before we move away from coal and when we do we'll probably be using Chinese-made solar panels rather than home-grown ones. 'We can never compete with China and India in making solar panels and windmills,' he says. 'Never will. Impossible.' Still, our common ground on energy efficiency means we are starting to get along well enough to joke around a bit.

'Maybe you will even show up on each other's Facebook pages now,' suggest Tony.

'I don't know, no, I don't have a Facebook page,' says Nick.

'I would Facebook friend you,' I offer.

'Sorry,' says Nick.

'You've got to convince him, Anna, that it's worth it,' says Tony.

'Clearly my next project,' I say. 'I can change your mind on Facebook.'

Jokes aside, I'm disappointed that Tony's research shows that one-quarter of Americans (those in the 'doubtful' and 'dismissive' categories) still don't accept the science of human-induced climate change. 'This issue is still, you know, working its way through the American mind. Many people still don't understand a lot about it,' says Tony.

He tells us that 'there's no question' in his mind that the conflict over the science will eventually be resolved. 'At the end, nature bats last,' he says. 'And we're going to see the implications and the consequences of climate change increasingly over coming years. At some point we will act . . .'

But Tony isn't overly optimistic. 'The question is, is it before it's too late?' he asks. 'Unfortunately this is one of those issues where literally an ounce of prevention is worth a pound of cure.'

Chapter Fifteen

BJØRN AGAIN

NICK AND I ARE SITTING OUT THE FRONT OF AN ORGANIC VEGAN RAW food café on a busy New York city street near Union Square. The person who selected the restaurant is one of Nick's picks: the Danish author Bjørn Lomborg. Over the past decade, Lomborg has made a name for himself as a global brand. He is known as 'the sceptical environmentalist'.

Kate drives off in our hired van to pick up a bicycle from the other side of town. Max wants set-up shots of Bjørn dismounting from the bike outside the café. It will look as if Bjørn cycled to our meeting, even though he actually didn't, and it's designed to make Bjørn look like an environmentalist. I'm staggered. His participation in this moment for me sums up pretty much everything about the man we're soon to meet.

The self-styled 'sceptical environmentalist' is a curious creature. Conservatives had once tried combating environmentalism from a position of economics ('it costs too much to save the planet'). They found it didn't work. Most people thought saving the planet was worth it.

So a space opened up for a new type of intellectual: the anti-environmentalist environmentalist. Smart conservatives began to cultivate and support spokespeople who pretended to be motivated by saving the planet but in practice argued *against* policies like cutting carbon pollution. Everything I've heard about Lomborg seems to put him into this category.

Former US president George W Bush was a master at this type of framing. He cleverly used words to influence the way his policies were perceived, even if the policy had the opposite effect of the message. Hence, Bush dubbed an environmentally destructive piece of law the Healthy Forests Restoration Act. His controversial education policy that, among other things, encouraged the teaching of creationism in schools was anointed with the caring-sounding name 'No Child Left Behind'.

Lomborg seems to be a very clever political operator with the same tendencies. He spends much of his time claiming to be a 'data-driven environmentalist' in mainstream media outlets around the world. In reality he's a *persona non grata* to most environmentalists and climate scientists who've taken a closer look at the policies he argues for.

Bjørn used to be a high-profile climate sceptic. In 1998 he wrote that 'the Greenhouse Effect is highly doubtful'. He'd also said, 'the greenhouse effect is coming. Maybe. This is the newest of our myths, and still the most difficult to assess. But we know that the theory is based on very problematic models.' Back then he argued that sun spots were a more convincing cause of climate change than carbon pollution.

Then Bjørn shifted into a second phase, where he accepted the basic science of man-made climate change, but downplayed its risks and significance. In countless media interviews and speeches, he scoffed at the potential dangers of climate change as outlined by advocates such as Al Gore. He argued that attempts to prevent climate change would be too expensive and ineffectual. In particular, he loved to pose a dichotomy between addressing climate change and tackling 'more urgent' and cost-effective issues like HIV/AIDs.

In 2006 Lomborg appeared in a Fox News television special (*Global Warming: The Debate Continues*) that tried to paint a picture of scientific disagreement. On the show he claimed that climate change was not an imminent threat. 'The data, the facts tell you that many, many things are moving in the right direction,' he said. All of this provided convenient cover for politicians like Nick Minchin who oppose taking action on climate change.

Now, Bjørn seems to have entered a third phase: not only does he accept the science of climate change, he's also accepted that it's an urgent problem. In August 2010 the *Guardian* reported that 'the world's most high-profile climate change sceptic is to declare that global warming is "undoubtedly one of the chief concerns facing the world today" and "a challenge humanity must confront", in an apparent U-turn that will give a huge boost to the embattled environmental lobby.' Lomborg's latest book calls for tens of billions of dollars a year to be invested in renewable energy research and geo-engineering options. But while he accepts human emissions of carbon pollution are the cause of climate change, he still isn't ready to endorse a price on carbon pollution of more than $7 per tonne. He prefers direct investment in renewable energy research, hoping that this will lead to cheaper technology and therefore in time the phasing out of fossil fuels.

•

Bjørn appears for our meeting dressed in his trademark uniform of a black T-shirt and blue jeans. He's not on the bicycle; Max will film those shots at the end. He looks friendly and down to earth, with shaggy blond—almost bright yellow—hair. Many articles describe him as charismatic but I don't sense it today. He is, however, very talkative. He has the kind of relaxed confidence that you feel around people who are very sure of themselves. Within the first few minutes of our conversation it's clear he's an expert media performer. He'll have no problem getting his talking points across. Many times.

'There's a lot of scare in the climate conversation,' are Bjørn's first words. Well, yeah, I think. When you're 28 years old and staring down the barrel of a future irrevocably altered by climate change, it *is* pretty scary.

'I think we need to roll back and start talking about how are we going to make smart solutions to climate change,' he continues. 'Rather than the feel good, very expensive ones that haven't worked for twenty years.'

I ask him about the science and he concedes: 'Global warming is happening. It is man-made. It is an important problem.' I wait for the 'but' and Bjorn doesn't disappoint. 'But it is also very often exaggerated,' he says. He argues that this 'exaggeration' doesn't work. 'Because it only makes people think . . . I have to switch my light bulb or buy a Prius,' he adds. 'That's all good, but the real answer is to change the energy system. So that green energy becomes so cheap everyone will want it.'

I agree with him that we have to change the world's energy system. We need price signals to make clean energy cheaper. A price on carbon pollution, for example, gives business a financial reason to clean up its act. It also generates revenue that can be invested in developing renewable energy infrastructure, which is what Bjørn is passionate about.

The first step is to remove the subsidies that make fossil fuels so cheap (between $A6 billion and $A9 billion annually in Australia and US$409 billion globally in 2010). Only then will renewable energy be on a level playing field. The International Energy Agency (IEA) condemned fossil fuel subsidies as 'state aid' and 'a significant economic liability' in its 2011 *World Energy Outlook*. It stated that eliminating fossil fuel consumption subsidies by 2020 would cut global energy demand by four per cent and considerably reduce carbon emissions growth.

Bjørn talks about making clean energy cheaper but he doesn't mention fossil fuel subsidies, only research and development. He accepts that climate change is happening. He accepts that it's caused by human emissions of carbon pollution. But instead of policies that directly target this cause, he proposes a more circuitous route. He doesn't think we should focus on regulating or cutting carbon pollution. 'Even if you do it [cut carbon emissions] effectively it's going to cost a lot,' he says. 'And it's probably not going to happen. And even if it does it won't do very much.'

Clean energy today is plentiful and affordable. We *have* the technology to solve climate change. But Bjørn doesn't seem to believe in rolling it out yet. He thinks it still costs too much. He would prefer we invest in research and development until some unspecified time in the future where clean energy becomes cheaper. 'If you spend money on research and development into green energy,' he says, 'make sure it becomes cheaper than fossil fuels over the next twenty to forty years—we've solved global warming. If we could make solar panels cheaper than fossil fuels by say 2030 or 2040 we'd be done.'

There's just one problem with Bjørn's argument: we don't have until 2030 or 2040! The IEA and the world's scientific community argue that the world must stabilise and then reduce emissions by 2017. To reiterate the words of the IEA's chief economist: 'If, as of 2017, we do not see a major way of clean efficient technologies then the door to [stabilising emissions at] 2 degrees will be closed and will be closed forever.'

That gives us five, not twenty-five, years. We must meet this 2017 deadline, says the IEA's chief economist Dr Fatih Birol, if we're to avoid what scientists refer to as 'dangerous climate change', the level where the climate passes tipping points.

There's a simple reason for the tight 2017 deadline. The world has already used up eighty per cent of the emissions 'budget' required to stay under the 2-degree warming threshold. Soon the world will have built so many new fossil fuel power stations that enough emissions will have been pumped into the sky to get us over that threshold. And once new fossil fuel infrastructure has been built, it will be in use for decades. Irreversible climate change will be locked in.

So Bjørn is right that we need to change our energy system. But we can't simply sit back and watch as enough new coal-fired power stations are built that will ensure we pass the climate's tipping points. Even if scientists get billions of dollars to research and develop new clean energy technologies, by the time these come into use it will probably be too late. We *already have* renewable technologies available to wean the world off fossil fuels. Yes, we need clean energy research, but we also need implementation, and we can't wait.

•

Most of our conversation with Bjørn Lomborg consists of him repeating his point about investing in clean energy research and development over and over in tidy sound bites. I try to steer him into talking about why he accepts the science of climate change. Nick tries to steer him away from the science.

Here's an example of our conversation:

NICK: *And we've met people like Dick Lindzen and I suppose I am more inclined to his sort of view of the world. We have had an interesting debate about the science per se. But what I think is interesting in your case is putting, in a sense putting that to one side . . . is what we are looking at dangerous? And to the extent that there is a problem, how do we deal with it?*

ME: *I just think it is such a great opportunity that we have Bjørn here. Someone who does accept that climate change is caused by humans. Maybe there is something that he could explain to you about why he's come to that view that you might listen to.*

NICK: *Yeah, although the greater value in Bjørn is the issue of whether the warming that's forecast is dangerous and to the extent that there is a problem what is the best way of dealing with it?*

ME: *And why it's happening, because that obviously . . .*

BJORN: *Well, let me jump, let me try and jump in and maybe make some sort of . . .*

NICK: *We're talking to scientists about that, we're having a separate debate with scientists about that . . . what I think Bjørn brings to the table is his extraordinary research that he has done into the question of the threat and the extent to which and how we deal with it.*

In the end, Bjørn and Nick avoid the question about science so much that Max has to intervene. 'I'm going to have to grab a quick wide shot before we move on,' he says, towards the end of the interview, 'but before we do that there is one question that Anna has brought up. And I think we should sort of answer it directly, which was to ask Bjørn what you think about the science.'

'Sure,' says Bjorn. 'I'm a social scientist so I simply listen to the smartest climate scientists that we have on the planet. They all say, even very sceptical ones, more CO_2 in the atmosphere means higher temperatures. So I believe there is a problem. It is human caused. I think it is fairly obvious and straightforward.' As always, he adds a 'but'. 'But I think the real issue is . . . how should we tackle it and how should we be communicating it.'

I try to press him further on the issue of how he came to that view. 'All of us are non-scientists,' I say. 'What kind of process do you go through when you do have people like Dick Lindzen out there—how do you judge all of this information?'

'Well I think that is why we have the UN Climate Panel, the IPCC,' says Bjørn. 'They tell us this is the best evidence when we collect everything. You know, that's why I think we need to have some sort of broad acceptance. That we accept all of what they say . . .'

'You trust the IPCC?' asks Nick incredulously.

'I think, you know, the IPCC is by no means a perfect process,' says Bjørn. 'But it's the best we have. So I think we gotta say, they're telling us there is a problem, and we gotta, you know, start thinking about how we are going to act.'

But Bjørn still believes that some of the predictions made by scientists should be challenged. Nick leads him to talk about one of them. 'You make the point about deaths . . .' says Nick.

'Most people die from cold and not heat,' says Bjørn. 'So you'll actually have . . . fewer deaths overall.'

This doesn't sound right to me. I don't have the figures on hand so after our meeting I look into it further.

This doesn't sound right to me. I am aware of a *Newsweek* article by journalist Sharon Begley that digs deeper into this particular claim. Begley's article discusses a book by media critic Howard Friel called *The Lomborg Deception*. As the title suggests, Friel dissected Lomborg's work and found it lacking academic integrity.

It was quite a process that led Friel to this conclusion: he painstakingly checked every citation in Lomborg's 2007 book, *Cool It* and found that, while the book appears well-referenced because of the sheer volume of footnotes, many of Lomborg's claims (including the one he's just told Nick and me about cold deaths) are essentially unreferenced and potentially untrue.

Begley asked Friel specifically about Lomborg's claim that global warming will avert more deaths than it will cause due to fewer people dying of cold. Lomborg cited five sources as references for this claim—but Friel checked them and found three reached the opposite conclusion! (Lomborg told Begley he included them simply to criticise them. However, it's very unusual to list articles in a footnote without noting that they're there to be criticised.)

For the remaining two references, Lomborg told Begley that 'there is no question that they support my point. Indeed their support is so explicit that I am at a loss to see how Friel could have construed it otherwise'. However, Friel found that one of the studies states that 850,000 deaths from cold will be averted in a warmer world, significantly fewer than the 1.4 million Lomborg claims. Furthermore, the study compared deaths in a colder world with deaths in a warmer world by only looking at six causes (cardiovascular disease, respiratory illness, diarrhoea and three tropical diseases) rather than a comprehensive comparison of all deaths. The other study examined estimates of death rates for people aged 65 and 74 only. Begley points out that this 'is hardly a full population-wide analysis'.

Finally, Lomborg cited a report from the World Health Organization (WHO) as a reference for his claim that cold kills 1.5 million people in Europe annually. But Friel found that the WHO report says nothing of the kind. When Begley asked him about this, Lomborg said he cited WHO 'solely to provide an estimate of Europe's population'. After hearing this, Begley notes: 'but as with other source notes, it [the reference] appears to support his controversial claim, not something as unobjectionable as Europe's population'.

After examining numerous examples like this, Friel concluded that 'this pattern of nonexistent footnoted support for assertions in the text was quite common'. He found that Lomborg often makes 'a highly substantive claim that, when you go to the footnotes, is not supported'. The problems with his work appeared to be so grave that Friel concluded Lomborg was 'a performance artist disguised as an academic'.

As a side note, I remember later that Working Group II of the IPCC released a report in April 2007 that showed, among other things, that fewer deaths from cold exposure 'will be outweighed by the negative health effects of rising temperatures world-wide, especially in developing countries'.

But this isn't to say that there's nothing of value in what Bjørn has to say. He agrees that climate change is happening and is caused by humans. I agree with him that we need to change the energy system and make clean energy cheaper than fossil fuels on a global scale. We just differ on how to get there. I think his strategy of just investing in research and development is too risky given the timeframe. He thinks the conventional strategy of regulating carbon pollution is too expensive. But at least we're arguing about the solutions, rather than still being stuck on the science.

•

After filming the final shots of Bjørn dismounting from the bicycle that's been driven over from the other side of town, we say goodbye.

Then we have a rare afternoon with no filming. I quickly pack my bags so they're ready for tomorrow's long-haul flight to London. Then I catch the subway to Manhattan's Upper West Side to visit two friends of mine from Australia and their new baby, Kate. I stop at the Grand Central Station market to pick up some fruit and veggies on the way. It's so great to see my friends again and I'm relieved they don't look too sleep-deprived. When I meet Kate, I stare in awe. She's tiny. She's perfect. I cradle her in my arms and she wraps her teensy hand around my finger. She is the highlight of this whole journey so far.

After a few hours I say goodbye and start to walk back to the hotel. It's humid, with dark clouds above. The sun is sinking low in the sky. I breathe in deeply, all the distinctive smells and sounds of New York rushing in. Soon it starts to rain— big, fat, warm drops. I can't help but grin. I love New York in the rain just as much as I love it in the sunshine. I'd love to stay here in this city that feels like it never sleeps. But instead, tomorrow, I'll board a plane to London. It will be the last leg of the trip.

As much as I love the States, I am looking forward to being on the continent that's leading the world in action on climate change. I can forgive Europe for exporting Bjørn Lomborg from Denmark. They've more than made up for it by charging into a cleaner future, with many smart steps to reduce carbon pollution. Denmark, for example, generates more of its electricity from wind than any other country in the world. Wind power accounted for around one quarter of its total electricity in 2011, with the government planning to increase this to 52 per cent by 2020. Europe recognises the need to act now to reduce carbon pollution. Bjørn believes we must wait until some unspecified time in the future when clean energy is cheaper, but the majority of Europe's governments and businesses disagree with him. They understand that the world can't wait.

Chapter Sixteen

THE CONSERVE IN CONSERVATIVE

I'M BARELY AWAKE AS WE FLY OUT OF NEW YORK AT 5.30 AM. SEVEN sleepless hours later, we touch down at London's Heathrow airport. I look like death, having been seated directly behind a baby I've christened 'Lungs' in recognition of his favourite organ. My energy levels have been drained more and more each day with the relentless demands of our filming schedule. Multiple interviewees each day, difficult conversations with Nick, finding smart ways of communicating complex science to a television audience and the filmed debriefs in the evenings are all taking their toll. Combined with lack of sleep and constant travel, I'm not feeling great.

The last thing I want right now is for this moment to be filmed for national television. And Max knows this, partly because I told him and partly because, when I see him with the camera in baggage collection, I do a swift about-turn and walk in the opposite direction. This doesn't deter him. He wants 'airport shots' to mark the transitions between our locations.

I feel like a giraffe trying to take evasive action against a particularly persistent hunter. I'm too tall. I try to blend into the crowds and hunker down in badly lit corners because I know that without good lighting, the shots won't be useable. Ultimately my subterfuge fails; Max tells me he'll be filming as I walk out the terminal door. There's nothing I can do.

The next morning I wake up in a jet-lagged fog to a loud, insistent noise. As I stick out one arm from under the blanket and fumble around in the dark to turn off my alarm, I can't for the life of me recall who we're meeting today. Is it another one of Nick's choices?

Then I smile, because I remember. Today is going to be a good day. We're scheduled to meet Zac Goldsmith, former editor of *The Ecologist* magazine and current member of the British Parliament for Richmond Park. What's more, he's an up and coming leader in the Conservative Party.

Even though they have the same ideological roots and philosophy as the Australian Liberal Party, the UK Conservatives have taken a very different approach to climate change. Instead of rejecting the science, they take it just as seriously as the other British political parties. Their platform at the last election included a raft of credible initiatives around reducing carbon pollution and playing a leading role in the international climate negotiations.

Nick, Kate and I take a cab to Westminster Palace, the impressive building that serves as the meeting place for both houses of the British Parliament. It's unseasonably warm for early October and the newspapers are all calling it an autumn heat wave. Yesterday was 30 degrees Celsius—the hottest day in October since London's records began in the late 1800s. A few days before, the city had experienced the warmest end of September for more than a century.

We all assemble with the lighting and camera gear. The area we'd planned to film in is blocked off today because of an official function. Instead, Max directs Nick and me to stand in front of a statue of

Oliver Cromwell. He gets me to read out loud an introduction he's written, which says I've brought Nick here to a statue of the 'father of British democracy'. It's a mistake. Nick shakes his head and correctly points out that Cromwell was actually a dictator. Great, I think. It's going to look like I'm advocating some kind of green dictatorship.

After finishing the setup shots, a tall figure strides around the corner of the cobblestone street on which we're standing. Handsome, silver-haired and immaculate in his black suit, it's Zac Goldsmith. He's smiling hello, while carrying on a vigorous conversation on his mobile phone.

I'm excited about this meeting. Sir David Attenborough had been keen to be our British interviewee, but was unavailable as he was filming in Borneo. So I'd thought hard about who else in London to choose with the potential to change Nick's mind. Britain punches above its weight in both climate science and climate solutions, so an almost overwhelming number of potential spokespeople came to mind.

I focused on finding someone who was able to speak to Nick's core values of economic freedom and small government. Ideally, I wanted someone involved with the governing Conservative Party. I thought Nick might listen more to one of 'his people'.

Zac Goldsmith quickly made his way to the top of the long list. Seeing how professional and competent he looks as he talks on the phone, I feel I made the right choice. I hope he can make some headway with Nick today by building on their common ground.

At the very least, I hope he covers three key points: that the rest of the world is acting on climate change; that this is creating rather than destroying economic opportunities; and that the future of Conservatism as a political movement lies in embracing rather than rejecting these economic opportunities. I also hope that Zac, as a fellow Conservative politician, can persuade Nick that his fears about big-government interference in the economy shouldn't get in the way of accepting the science.

After Zac's call is finished, Max arranges the three of us on the lawn in front of Westminster. Zac begins by asking Nick if he can clarify his position on climate change. 'Your view is that . . . it's happening but it's not man-made?' he asks in his polished British accent.

'Well, my view would be that we had, you know, this warming up until about '98, 2000—human emissions of CO_2 probably made some contribution to that,' says Nick. I smile and try not to cheer out loud. Nick just said, for the first time on camera, that he accepts carbon pollution 'probably' plays some role in causing climate change!

What a breakthrough. This statement undermines some of Nick's former statements, showing he has accepted at least part of the basic science of human-induced climate change. Up against a rational, conservative leader like Zac, Nick's concession demonstrates how obsolete it would be on the modern world stage to completely deny the causal role of carbon pollution in climate change.

'The extent of that [the contribution of CO_2 to warming], I think, is still a matter of substantial debate,' Nick quickly adds. 'And then there's a question of is it dangerous—the warming that's occurred, is it likely to continue, and what policy response should there be?'

From Zac's position, as an MP in a political system with bipartisan acceptance of the need to act on climate change, there is no longer a debate about the science.

'In as much as science ever allows a consensus, we're not far away from that in terms of dealing with this issue,' Zac says politely. 'And all the national science academies of all the G8 countries and many more—I think thirty-two national science academies—have co-signed statements saying this is an issue, we're behind it and we're going to have to take action . . . I think in terms of basic risk assessment we'd be pretty brave to discard all of that and say they're wrong and we're going to continue with business as usual.'

Zac is obviously used to talking to the media. He uses clear analogies about the nature of risk in public policy. 'I think it was Dick Cheney [former US vice president under George W Bush] who is quoted as having said that if there was a one per cent chance of Al Qaeda accessing nuclear bombs he would treat it from a policy point of view as a certainty,' he says. 'We've got more than a one per cent chance of climate change as a real issue. And the upside of taking action . . . is that most of the things we would do to tackle climate change are things we would have to do irrespective of climate change. So for me it's a no-brainer.'

'Is your motivation that you genuinely believe the world is on course for very dangerous global warming and that we've got to do as much as possible to reduce our emissions to avoid that dangerous warming?' asks Nick.

'It's more than that but I suppose that's a starting point,' replies Zac. 'The starting point is that I'm convinced by the science that I've read and I'm convinced by what looks more and more like a consensus . . . the bulk of science tells us that we're facing a really serious problem and that is really a major concern to me.'

Zac also believes that reducing emissions is essential for economic reasons. 'There's a recognition from all the parties [in the UK] that we need to head towards a low-carbon economy,' he says. In fact, it's the economy that dominates the bulk of our conversation. I am hoping Zac will talk more about the economic benefits of acting on climate change. I'd love him to rebut the argument Bjørn Lomborg put, that cutting carbon pollution simply costs too much until new technology is invented.

I ask Zac a question along these lines. 'Nick introduced me to two people recently—Bjørn Lomborg and Marc Morano—who both say that it's economically irresponsible to act now, that we should wait until some point in the future,' I say. 'What's your position on that?'

'Well I've also debated Bjørn Lomborg quite a few times,' says Zac. 'Over the years his position has shifted a fair bit. When I first debated him it [climate change] wasn't happening. The second time I debated him it was happening but we weren't causing it. The third time I debated him it is happening, we are behind it, but it's not cost effective to deal with it. And I hope the next time I debate him he will say that it is cost effective because we have no choice.'

Zac talks about some of the economic reasons to act now on climate change. 'Some of the figures used by climate sceptics, I think, confuse cost with investment,' he says. 'So if you invest 100 units in making your home more energy efficient, for example, in real terms that's an *investment*. You're going to get paid back within four or five or six years based on technology which exists today.'

Zac talks about a number of programs the UK Conservative government has implemented to reduce carbon pollution, making a strong case for putting the 'conserve' back into Conservatism. 'I obviously belong to the Conservative Party,' he says. 'And I think if it's a given that our defining challenge now is harmonising the market with the environment, that means finding market solutions. I don't think historically there has ever been a political organisation in this country with as good an understanding of market mechanisms as the Conservative Party.'

He speaks about the UK's Climate Bill, a major piece of climate legislation which set a target in law of cutting the nation's carbon emissions by eighty per cent by 2050. It was supported by all the mainstream political parties. 'Its biggest advocates were in my party, the Conservative Party,' says Zac. 'This was a good way of kind of kick-starting the economy.' He points out that the passing of the Climate Bill 'coincided with the beginnings of a vast economic gloom . . . there's a recognition that when we emerge from the mess that we're in at the moment we ought to try to do so with an economy that might actually last as opposed to one which is based on waste.'

I wish we could bring Zac to meet the entire Australian Liberal Party. It's so refreshing to hear the economic downturn used as a reason to *clean up* our carbon pollution, rather than an excuse to delay action.

Zac talks about the smart steps the UK government is taking on energy efficiency. 'We're doing energy market reforms . . . that will put energy saved on a par with energy generated,' he explains. This means that if you're an energy company, you will earn just as much by investing in energy efficiency . . . as you would providing energy itself.'

It's clear that Zac is excited by the changes this will bring to the UK economy. 'It will completely transform the dynamic of the economy. You have an immediate incentive then, a pressure, a market pressure on companies to deliver [energy] savings and not just energy.'

This is the crux of the debate I'm hoping Nick and Zac will have. By now it seems clear to me that Nick rejects climate science because he's scared the solutions will involve too much government regulation, which goes against his core values. I really want Nick and

Zac to have a productive discussion about the role of government regulation when it comes to climate change.

'I think it's important to respect or acknowledge that there is no power on earth as powerful as the market—other than nature itself,' says Zac. 'As a Conservative I think we'd probably be in absolute agreement on that. But the market has blind spots—and in many ways the environment is the biggest blind spot. So there is a need, in my view, for intelligent intervention. That's where conservatism has moved in this country.'

'It doesn't mean bigger government,' says Zac. 'I believe in small government . . . But it does require some intervention.' He tells us 'even Margaret Thatcher said "never call me laissez-faire". The government has to have the courage and the strength to do what only government can do. In terms of righting some of the wrongs in relation to the environment, regulation by government is absolutely key.'

Zac gives an example of the need for government regulation when it comes to over- fishing. 'We've wiped out almost all of the world's great fisheries. Fifteen of the world's seventeen fisheries are gone or about to go,' he says matter-of-factly. 'Without some kind of intervention—policy intervention with teeth—we are going to exhaust the world's oceans and that's going to have massive implications for future generations.'

'Society's got to regulate that,' says Nick. It sounds like he's in agreement when it comes to regulating fishing to prevent fisheries collapse. But can't he see the obvious parallel with regulating carbon pollution to prevent climate collapse?

'So there is a role of regulation,' says Zac. 'There's a role for intervention but it's got to be done carefully.'

'It's got to be—I mean from the Australian point of view I come back to this point—it's got to be hooked into the reality of the nature of the economy we've built after 200 years,' says Nick. 'It's quite unlike most European economies . . . you don't just switch that overnight. So, you know, and I accept we discussed this, that ultimately one day . . .'

'But what we're saying is we have to start the transition *now*,' I say. 'We can't keep putting it off, Nick. What's going to happen to the Australian economy if you have countries like the UK and Europe and China acting and we're being left behind?'

Nick doesn't agree. 'Governments are absolutely, absolutely bloody useless at picking winners . . . so, you know, a lot of these things have got to be left to natural economic forces,' he says. But what does he mean, *natural* economic forces? The economy was created by humans, and the way it operates is governed by human laws. In my mind, there's nothing natural about it.

'Now, there is a role in regulating,' concedes Nick. 'The question is, do you regulate the emission of CO_2? That's the particular thing we're talking about.'

Zac tries to convince Nick that even if he doesn't yet accept that regulating carbon pollution should be done because of climate change, Nick should support it anyway because of the opportunities it creates for a clean energy economy.

'The reason why we don't have . . . any significant opposition to the policies [in the UK] is because people can see that we are at the cusp now of a massive industrial revolution,' says Zac. I had told Zac about Nick's comments in his *Four* Corners interviews, and he obviously remembers them: 'I know you've talked about climate change as being an excuse to de-industrialise society,' he says. 'We see it as an opportunity to *re-industrialise* society, but in a different way. We'll have a different type of economy. A low-carbon economy. That transition is monumental.'

I turn to Nick. 'Can I just ask—do you still think climate change advocates are essentially arguing to de-industrialise Western society?' I wonder if, after spending three weeks with me, I've at least managed to show him that he's misguided on this point.

'That was a quote which has been much misrepresented,' says Nick. 'I said that this cause attracts a whole lot of people but it particularly attracts deep green, you know, activists and environmentalists who are using this issue because they genuinely believe, and I put no proportion on this, but I was obviously talking about a component of those who adopt this cause because they do want it de-industrialised. There's no doubt about that. That's incontrovertible and it wasn't to slander everybody involved in this cause.'

This comment is insightful. It reveals Nick's fear that climate change could be used as a Trojan horse to increase government regulation and damage the economy by anti-capitalists. But in focusing on these fears, Nick is missing the big picture of the much greater damage that unchecked climate change will inflict on the economy. He's missing the forest for the trees. It's like living in a bushfire-prone regional area and refusing to put a water tank next to your house because you think it's ugly and might reduce the value of your property, without realising that without the water tank your whole house is at risk of burning down.

Nick seems to fear that accepting climate change science will lead to government-owned renewable energy infrastructure. Nick supports privatising public infrastructure, not creating more of it. (In his final speech before he left Parliament Nick said, 'I failed to achieve the sale of a government-owned private health insurance company called Medibank Private. I dare not even mention what else I would like to have sold.')

But Zac tries to explain that acting to reduce carbon pollution doesn't necessarily mean heavy government intervention of the sort Nick fears. 'I agree governments never pick winners correctly,' says Zac. 'I think it is nevertheless important that they create a stimuli, they create a framework which the market can then respond to. That's not the same as picking winners.'

Zac refers to Britain's landfill tax to give Nick an example. 'The most wasteful sector in this country is the construction sector,' he says. 'The landfill tax, when it was introduced, everyone said it's impossible, you're just going to be hammering this sector. They can't do anything about it therefore it's no point imposing a green tax.'

This is similar to the arguments made by opponents of Australia's price on pollution: that the 500 or so biggest polluting companies affected by the scheme won't be *able* to reduce energy or switch to renewables. Polluting companies argue that there's no room to innovate to reduce emissions. They will just continue business as usual, they say, and either go broke paying the levy or pass the price on to consumers. But despite similar scare-mongering, this isn't what occurred when Britain introduced its landfill tax.

'As it happens, the construction sector has cleaned up faster than any other sector in this country,' says Zac. 'There's a company quite near here which is the second biggest construction company in the country . . . and because they fear that the landfill tax is going to increase again they have now taken steps to become a zero waste company,' he recounts. He sounds very proud of their efforts. 'They've succeeded, and not only are they now avoiding any future rise in landfill tax . . . they're already saving an absolute fortune by avoiding the existing landfill tax. And I think that fairly crude but nevertheless unavoidable stimulus provided by government is quite often what's needed.'

'As long as there are cost-effective alternatives for them,' says Nick. He believes the arguments made by some energy inefficient companies in Australia, like aluminium smelters, that they simply can't reduce their levels of carbon pollution. He refuses to countenance that we might be on the cusp of a new industrial revolution. I doubt Nick would have argued strongly in favour of the horse and cart industry just before motor vehicles became commonplace. I *know* he didn't protest in Parliament on behalf of the film camera industry when Australians started switching to digital cameras. The reality is that industries change over time. In Australia, we're fast coming to end of the age where it's environmentally or socially acceptable to continue using eighteenth-century coal-fired power.

'You never raise a regulatory burden beyond what is possible, absolutely right,' says Zac. But Zac knows, like I do, that the world *does* have the existing technology and know-how to eliminate the wasteful practices that have created climate change and other environmental problems. 'I mean the reality is, if you took best practice today in every field and you made it the norm tomorrow, we wouldn't be having this discussion. We'd be there,' he says. 'We know that the market is capable of thinking up extraordinary solutions . . . but they normally happen as a result of market opportunity and market stimulus and that's occasionally where government comes in.'

'And that's the thing with [putting a] price on pollution,' I add. 'It gives business a reason to clean up their act that they wouldn't have otherwise. It's quite simple.'

'No, no, no,' says Nick firmly. 'If you're saying that we're going to tax the emissions of CO_2 ... you must find an alternative—there's got to be a cost-effective alternative.'

'Yeah, you must reduce your emissions,' I say. That is the alternative. You can do it through energy efficiency, or from switching from polluting energy to clean energy. I seem to have a much more optimistic view of the ability of business to adapt than Nick does. 'Australia can be powered by a hundred per cent renewables within a decade,' I tell him.

'Within a decade?' asks Nick incredulously.

'Yeah, it's just a matter of when we start,' I say. 'We need to start as soon as possible. We're already behind countries like the UK.'

'That's too—that staggers me—for me to even think about,' replies Nick. He seems shocked by the idea.

Perhaps this is the key generational difference between us. My motto is: 'Those who say it can't be done should get out of the way of those already doing it.' Some, like Nick, call me naive for taking the position that Australia can be powered with a hundred per cent renewable energy. But not only is this vision backed up by solid research and a clear blueprint, I simply believe that we have no choice. Given the scale of the climate change challenge, my other motto is: 'We have to demand the impossible to avoid the unimaginable.' I recognise that for Nick, a hundred per cent renewable energy in a decade might seem impossible. But for the sake of this generation of young people and the millions of Australians to come, we must at least try.

Zac takes the middle ground, as you would expect from a Conservative Party member. 'I think realistically we're not going to transfer immediately to wind, solar, energy efficiency and tidal—all the kinds of power sources which we are going to need,' he says. 'If you ask for medium to long term, I hope we will be a hundred per cent renewable. I believe that is an inevitability at some point but we're not going to get there immediately.'

But Zac does believe that the world can move a lot faster than it's currently moving, with the right policy settings in place. 'I'll just give you one example as to how quickly things can happen and what policy can achieve,' he says. 'There's one town in Germany called Freiburg, which is about 100,000 people, which produces more solar energy than the whole of the UK combined. That's because the government introduced something called the feed-in tariff, which rewards people for putting solar panels on the roof.'

'I think it's essential that we find a way of turning carbon into a liability,' says Zac. 'If we don't do that we're not going to see any kind of real incentive in businesses to reduce the amount of carbon they emit.'

All this talk of creating new opportunities and jobs in clean energy leads Nick to put forward the Liberal Party's position on climate action leading to job losses. For me, it's an old way of thinking. 'People forget that you might be creating jobs, but you're also *destroying* jobs at the same time,' he says.

Zac disagrees. 'PriceWaterhouse—one of the biggest accounting firms here—was looking at the downside of green policies,' he says. 'There are future potential job losses, but they're unable to identify any job losses at all at the moment caused by green policies. Whereas it's not difficult to point to jobs that have been created on the back of green policies. Obviously if you make carbon a liability you create an opportunity there for people to develop things at low carbon and you create job opportunities.'

Zac's point is important. Often when polluting industries talk about 'job losses' they're talking about new jobs (that would have been created without a price on carbon pollution) not being created, rather than the axing of existing jobs. For example, in 2009 the Minerals Council of Australia made a well-publicised claim that the then-planned Emissions Trading Scheme would result in 23,510 fewer jobs in Australian mining over a ten-year period than would otherwise be the case. Bruce Chapman, the policy impact director at the Australian National University's Crawford School of Economics and Government, looked at this figure in more detail in a 2011 paper released by the Australia Institute. Chapman concluded:

> *The projected job losses from the ETS, particularly when considered over a 10 year time horizon, are in a statistical sense close to invisible with respect to employment and unemployment stocks, and trivial with respect to aggregate flows in the labour market. Also, it is apparently the case that with respect to mining sector employment, the projected losses are a very small proportion of overall inflows to and outflows from mining. Further, it seems to be the case that those leaving mining periods of growth are not then entering a protracted period, and more likely any period at all, of unemployment.*

Non-creation of, in Zac's words, 'future potential' jobs is very different to cutting existing jobs. Yet Australian newspapers did not pick up on this nuance. 'Miners say 23,000 jobs at risk,' screamed a headline in *The Australian* in May 2011 when the Minerals Council re-released their 2009 figures. Months later, Liberal Party senator Alan Eggleston still seemed to be misusing the figures, claiming during the climate change debate: 'The Minerals Council of Australia last year said that the carbon tax will cost some 23,000 jobs . . . I wonder how those people will feel about the ALP's actions today when there is no pay cheque to take home.'

In his report, Chapman also noted 'most economists would argue that any changes in the relative price of carbon-producing output must also be associated with offsetting *increases* in employment as a result of the higher level of activity in, for example, alternative energy production'. This echoes Zac's point about new jobs being created in emerging clean energy industries.

Towards the end of the interview, Nick tries to rebuke Zac about a letter sent by the British Conservative prime minister David Cameron to Australian prime minister Julia Gillard congratulating her on the passing of the carbon price.

'I should take this opportunity on camera to register the disappointment of all the Liberals I know in Australia at David Cameron's very unfortunate intervention in Australia by writing to our Labor prime minister applauding her on her carbon tax,' says Nick. 'I mean he should have known that we oppose it. The party with which he is in relationship opposed it,' says Nick.

'Look, I don't know the process that led to that,' says Zac, 'but it also shows that there is a recognition here within the Conservative Party that some issues are bigger than party. Where a government, whether it's left, right, far left, far right, centre, whatever, is taking the right step then it is appropriate for a leader of another country to recognise that.'

It's interesting watching Nick and Zac talk—they seem to represent the past and the future of conservative politics. Nick's insistence that a Conservative British prime minister shouldn't have congratulated an Australian Labor prime minister seems petulant and party-political. Zac seems statesmanlike in contrast.

'You will have a system of carbon taxation, and cap and trade, something will be introduced in Australia, almost inevitably,' says Zac. 'I would be willing to bet anything you want.'

'We're committed to repealing it,' says Nick.

'But at some point there will be a mechanism brought into Australia for good,' says Zac. 'It will happen. The question is, do you want that to happen now and allow yourselves to adjust to what is without a doubt the beginning of a transition to a new economy, or do you do it later on?' He pauses. 'I mean as a friendly adviser . . . I would really love to see your country engage now. Because I think you will benefit economically as well as environmentally.'

I think it's sad that Nick, a former finance minister, seems to have let his ideology blind him to an incredibly obvious global economic trend. The world's major economies are shifting to a low-carbon future. In 2009, renewable energy overtook fossil fuels in new investment for power generation. Wind, solar and other clean technologies attracted US$140 billion compared with US$110 billion for gas and coal-fired electricity generation. The biggest growth in renewables came from China and India. In 2011 the disparity grew further, with US$187 billion in renewable energy compared with US$157 billion for natural gas, oil and coal. Between 2004 and the end of 2011, investment in renewable energy, energy efficiency and smart-energy technologies surpassed US$1 trillion. 'The trillionth-dollar milestone shows that the world is not waiting for a deal on climate in order to start turning the super-tanker away from fossil fuels,' said the CEO of Bloomberg New Energy Finance, Michael Liebreich.

Zac tries to help Nick see this by talking about what the rest of the world is doing. He points out that China alone invested US$34 billion in clean energy last year. Closer to where we're currently standing, Scotland is set to produce almost a third of its electricity from renewable sources by the end of 2011. In 2011, £750 billion worth of Scottish renewable projects were switched on, with another £46 billion in the pipeline. The Scottish Department of Energy says the country is on track to meet its target of a hundred per cent green energy by 2020. A report published in the journal *Energy Policy* in 2011 showed that by 2030, the entire world could be powered by a hundred per cent renewable energy. It wouldn't be easy—we'd need to build 4 million wind turbines, nearly 2 billion solar photovoltaic systems and about 90,000 solar power plants—but at least we now know it's possible.

'From the point of view of children born today,' says Zac, 'Pressure on the world's resources . . . is going to be the defining factor. That is what will define their lives. So any companies that move now in terms of improving our ability to be efficient, and any countries that get into position where they are also more efficient with their resources—they're going to have a massive advantage. I think it's not just carbon, it's about resources generally.'

'Yeah, I'm with you on that,' says Nick.

'My view is that change is going to happen whether we like it or not,' Zac tells Nick. 'It's much better that it happens on our terms rather than being imposed upon us by external forces—whether economic, social or environmental.'

'It's really difficult,' continues Zac. 'But if we don't find a way of doing it somehow—and as I said before, I think Conservatives are in the best possible place to try and unpick that puzzle—if we don't then I think our future's going to be really very troubling and dark indeed.'

Zac looks at his watch. He's running late for the Conservative Party's annual conference in Manchester. As Leo takes off his microphone, he and Nick chat about their shared opposition to the European Union. They certainly have a lot in common.

Zac covered all the key points I'd hoped he would. He communicated that it's economically responsible to take the science seriously and act on it; that the rest of the world is acting; and that there's no need to be afraid that accepting climate science will lead to some kind of communist–authoritarian revolution. But he still didn't manage to change Nick's mind.

●

Later, Nick and I sit next to each other on one of London's new fleet of hydrogen buses and debrief. Trying to hold the camera steady, Max films us from the seat in front. As the bus snakes its way through the streets of London, Nick tells me that Australia's economy is different to the UK's and Europe's. Yes, *obviously* it's different, I think. We're one of the most polluting economies in the world, mostly because our energy system is still reliant on black and brown coal. In a 2008 study of OECD countries, Australia was the third most energy-hungry economy—with only Canada and the United States worse performers. The same report showed that Australia has major opportunities for energy savings in residential, commercial and manufacturing—up to 73 per cent, 70 per cent and 46 per cent respectively.

In addition to our energy-saving potential, Australia has huge untapped renewable energy resources, from large-scale distributed wind power, wave power, hot rock geothermal power, bioenergy, and of course solar photovoltaic and solar thermal. For a long time solar wasn't an option for base load power because the sun doesn't shine at night, but that problem has now been solved in Spain where they're using molten salts for storage. There are now solar thermal power plants in operation that send out electricity from the sun twenty-four hours a day.

In one day, the amount of sunlight reaching earth produces enough energy to satisfy the world's current energy demand for eight years. And Australia is one of the sunniest countries in the world. Yet while last year overcast Germany installed 7.5 gigawatts of photovoltaic power for the second year in a row, sun-blessed Australia has only a total capacity of just over 1.2 gigawatts. And a wind turbine installed in Australia will generate as much as twice the energy as the same turbine installed in Germany, due to our better wind resource. It seems mad for Australia to still be staring down coalmines rather than looking up at the biggest free sources of energy anyone could imagine—the sun and the wind.

'Imagine if politicians had performed the same foot-dragging maneuvers when it came to installing our telecommunications infrastructure,' points out Matthew Wright, the executive director of Beyond Zero Emissions. The first telegraph lines were immensely expensive. 'The time and money needed to support solar photovoltaic deployment across the country is nowhere near what the telecommunications sector needed. In Australia, it will take four to five years—not 100—to see cost-competitive solar energy enter the grid, driven by consumer demand and reductions in wholesale power prices.' Matthew's research shows that, in the near future, Australia could generate fifteen to twenty-five per cent of our electricity from solar photovoltaic power.

Because Australia's economy is so reliant on polluting coal, we're a long way behind the rest of the world when it comes to reducing carbon pollution. Luckily, we can apply the lessons learned by other countries like the United Kingdom, which have already made great strides towards sustainability. Once we take the first step, I have no doubt that Australians can unlock our potential for unlimited clean energy from the wind, sun and earth. One day that shining renewable energy future will arrive here, and I want to be part of making it happen.

Chapter Seventeen

MEDDLING IN THE MEDIA

THE NEXT MORNING DAWNS LOOKING MORE LIKE THE LONDON WE all know and love: grey, cold and rainy. Overnight, the weather has turned from a record-breaking autumn heat wave to a miserable, chilly mess. I'm already feeling the beginnings of a cold coming on. I hope Max won't make us do today's meeting outside. But I grab my red coat just in case as I run out the door of my hotel room. Counting on Max's benevolence has not been a strategy that's worked well so far.

True to form, Max tries to get Nick and me to do the requisite setup shot outside. Today's location is in front of Britain's national broadcaster, the BBC. It's windy and spitting rain. When the rain intensifies I draw the line. One setup shot that will result in less than three seconds of screen time is not worth me getting sick for the remainder of the trip.

Instead, we move inside to a Danish-themed café. I missed breakfast, so I'm starving. I eat some smoked fish salad and sip a cup of tea to warm me up while we're waiting for our next spokesperson, Christopher Booker. He's one of Nick's choices, and another curious character.

A 74-year-old journalist, Booker has built his reputation on being sceptical of mainstream science on a range of issues. He disputes the Darwinian explanation for evolution. He disputes the link between passive smoking and cancer. He believes white asbestos is safe, saying it is chemically identical to talcum powder. In 2007 Booker asserted that speed cameras had increased the accident rate where they were installed, citing government figures that in reality showed the opposite. He's also made bizarre claims about the European Union, stating that under EU rules people would only be allowed to bury dead pets after 'pressure cooking them at 130C for half an hour'. I'm still not sure where he got this idea.

So it's no real surprise that Booker is also a climate sceptic. His main claim to fame in this arena is his 2009 book, *The Real Global Warming Disaster.* It was described by *The Observer* as 'the definitive climate sceptics' manual', although the reviewer also noted that much of the book, 'including the central claim, is bunk'.

Booker comes into the café and sits at our table. He's keen to chat before we start filming. An elderly man with thin, wispy hair, spectacles and a penchant for a cigarette or two, Booker is the kind of journalist you can imagine in colonial India or Africa with a gin and tonic, and wearing a safari hat. He's an eccentric character sporting a sharp wit and a kind of gallant charm. An old-world gentleman, he lacks the nastiness of some of his fellow climate sceptics. When Nick asks if he knows Marc Morano, Booker says yes. 'Very noisy fellow,' he notes. They're on the same team, but worlds apart.

After our Danish lunch, we walk back to the BBC. It's stopped raining, but it's still icy cold. I pull my red coat tighter around me, wrap my arms around my shoulders, and wish I'd worn the thermals I'd purchased for Mauna Loa.

Booker calls the BBC the Ministry of Truth and claims it inspired George Orwell's vision of an authoritarian future. We're using the BBC as a backdrop because of Booker's allegation that the British media has created an unfounded scare in the public's mind about climate change. He believes that the foundations of climate science have been crumbling since 1997 or 1998. He is outraged that British sceptics are not allowed sufficient airtime in the media—especially on the BBC—to trumpet this.

'Anna will love this building,' says Booker as he looks at the BBC, 'because it is the citadel of global warming alarmism in the UK.' He sounds bitter. He worked there on and off in the 1960s and 1970s as a scriptwriter and documentary maker. 'About six or seven years ago,' he says, 'they decided that on this issue, since the overwhelming opinion of science all over the world was that global warming was real, very dangerous and the sceptics were just an insignificant little minority of people who can be ignored . . . that from now on policies should be totally, as it were, one-sided partisan on the issue.'

'So sceptics don't get a run?' asks Nick, shocked.

'That's why they must have all come to Australia,' I say. 'We've been having people like Lord Monckton coming on Australian television. Maybe that's why they left England.'

'I'm not going to talk about Lord Monckton,' says Booker. 'Who else? You haven't had any other sceptics from Britain? I mean there aren't that many,' he admits.

'Nigel Lawson, who was just recently—' says Nick.

'Nigel Lawson is obviously, as a very senior and respected politician, he's our leading political sceptic,' interrupts Booker. 'But you see, one of the great differences between Australia and the UK is that our political parties are all completely agreed.' Booker is referring to the bipartisan support for the science of climate change that Zac Goldberg spoke about yesterday. 'There is a consensus there . . . on the science,' he says. 'Whereas in Australia, thank heavens, in the name of democracy, you've still actually got a split and you've got an argument, you've got a big debate going.'

'That's partly thanks to Nick,' I say ruefully.

Booker argues that it's wrong for news agencies like the BBC not to give equal airtime to sceptics when reporting climate change. This is because, in his view, the science is not settled. According to him, climate change isn't caused by carbon pollution. 'Nobody actually knows what is deciding the climate,' he says. 'But if I err on one side or the other, I'm very firmly on the natural causes.'

Booker became a climate sceptic after he investigated wind farms for a regular column in the *Sunday Telegraph*. His research led him to look at the science behind climate change. 'What really first made me suspicious,' says Booker, 'was when I started to look at some of the methods used to promote the global warming belief theory, religion, call it what you will.'

'Do you still believe it's a religion?' I ask. 'Nick's used that word too.'

'Well it's common on our side of the debate,' says Booker.

'Yeah, we terrible sceptics,' adds Nick jokingly.

'We see this as a belief system, basically,' says Booker. 'And we then look at the evidence and we say hang on, there is a parting of the ways here. The thing that attracted my suspicion was looking at the hockey stick, the famous hockey stick graph.'

Booker thinks the hockey stick graph doesn't account for the so-called 'Medieval warm period'. Sceptics argue that this period was a time of higher temperatures globally. Most climate scientists disagree, since records show it was actually a localised warming event in Europe.

Booker believes that taking action to reduce emissions will destroy Europe's energy system and economy. 'The EU passed a huge policy document . . . which said we are committed, as it were, to lead the world on fighting global warming,' he says. 'So you should come and live in Europe, Anna, where we're all happy and in forty years time we won't have an economy left as a result.'

He talks derisively about the UK's target, enshrined in law, to cut carbon pollution by eighty per cent by 2050. 'Now, there is not one of the MPs who voted for that [who] could have begun to explain how we could do that without closing down a large part of our economy,' he says, 'because the technology is not there to provide us with electricity and ways to drive motor cars.'

This is an outright mistake: the technology is certainly there to meet the UK's target of cutting greenhouse emissions by eighty per cent by 2050. The UK government has just released a website with a calculator allowing the public to plot out multiple ways to reach and even exceed this target. The calculator allows the public to look at the options for reducing energy use through efficiency standards, and compare timelines and costs for transitioning the UK's electricity supply towards renewable energy.

'Now you tell me, Anna, how we're going to get that electricity?' asks Booker, without pausing to give me time to answer. 'I mean, you may say we don't need it. You may say we can live in the dark, we're quite happy. But the fact is that there is no science there yet or technology which can tell us how we're going to keep a modern electricity-dependent economy going just on useless windmills.'

I have to actively focus on my eyeballs to keep them from rolling. Does Booker have no idea about clean energy technology and how far it's come in providing base load power, or about electric vehicles powered by renewables?

'Windmills aren't useless,' I say. 'They generate power.'

'Yes, they're totally useless,' replies Booker. 'Little tiny bits of power occasionally when the wind goes up and down.'

Most people concerned about the economy aren't as quick as Booker to write off wind energy, and with good reason. In 2010 the wind sector globally had a turnover of €40 billion and employed 670,000 people. China and the United States are now the leading countries for wind generation, with China's National Development and Reform Commission planning for wind to generate 1000 gigawatts by 2050. This will generate about seventeen per cent of China's enormous electricity output. Bloomberg Investment suggests that even without a price on carbon pollution, this level of wind turbine manufacturing will have profound implications for the relative cost of wind energy. Bloomberg argues that it 'could even nudge down the cost of wind power enough for it to compete with coal and natural gas in the US and Europe'.

In Australia, the cost of wind has dropped dramatically, making it currently the most cost-effective source of renewable energy. Wind energy has the potential to power forty per cent of Australia's total grid-connected energy by 2020. According to researchers at Beyond Zero Emissions, this could be achieved through the construction of just 6400 new generation 7.5 megawatt wind turbines. And while some other estimates put the required number of turbines higher, even the highest are completely achievable. Home-grown wind energy currently costs around A$110 to produce per megawatt hour, although one project in Tasmania got the cost down to A$90 per megawatt hour. This compares with A$60 to A$120 for each megawatt of gas or coal generated to meet new demand. Once the carbon tax with its price on carbon pollution becomes operational and wind turbines are produced in larger numbers, wind will soon become cost-competitive with polluting energy.

According to Booker, though, the failures of wind farms are the least of our worries when compared with the way the public have been brainwashed by the media into believing climate science. Booker has a theory for why the world has been convinced man-made climate change is happening. He believes that in the late 1980s, due to the atomic bomb and the threat of nuclear war, humanity 'moved into a new phase where we actually were beginning to be conscious of the fact that man could influence the natural world around us'.

But then, Booker argues, 'it went off the rails . . . because of a limited little period with a resumption of a rising temperature trend which had been there in the early 20th century'. He thinks climate change was blown out of proportion. 'We panicked,' he says. 'We got carried away by thinking, ah, we can influence the world around us, nature, and this is the most supreme example of it . . . That's where the whole thing went off the rails in my view.'

'But climate science came before the '80s,' I point out to Booker. Human understanding of the fact that greenhouse gases trap heat dates back to John Tyndall's laboratory experiments in 1859, evolving even further in 1896 when Svante Arrhenius did the calculations that essentially predicted that the continued emissions of CO_2 would cause global climate change.

'Climate science!?' Booker exclaims. 'My dear Anna, forgive me, but climate science, which is a very imperfect science . . . a climate scientist to you, I suspect, is someone who agrees with the global warming theory.'

'No,' I say. 'It's someone who's studying human impacts on the global climate.'

'But in fact there are climate scientists on the other side,' says Booker. 'I mean Dick Lindzen of MIT is a very distinguished climate scientist but *he* doesn't agree with the global warming theory.'

I sigh. Against the thousands of climate scientists around the world, there's barely one climate sceptic scientist—the lonely Lindzen—with any credibility within the scientific community. And, frankly, after meeting him I wasn't very impressed. He provided no new evidence to support his previously interesting but now debunked theory about clouds. I'm at a loss as to why the sceptics keep putting him forward as a contemporary spokesperson when his scientific position has barely evolved in twenty years.

'We met Dick, we met him,' says Nick. 'It was terrific.'

Even though Lindzen's position is isolated, at least he is actually a working scientist. Booker's main job now seems to be professional denial of climate change. 'I'm writing this report about how the BBC has covered global warming,' he says. 'What comes across is what incredibly limited views their environmental correspondents have got . . . what they [the BBC] believe is press releases, IPCC reports and what a particular little group of scientists are telling them,' says Booker.

'But this *particular little group* encompasses every single national academy of science; it encompasses thousands of scientists around the world,' I say. 'How can you characterise the vast majority of science as a little bubble?'

'I'll tell you something about science,' says Booker. 'It moves on. And at just the point where everyone has finally worked out that, yes, there *is* such a thing as the law of gravity, someone else has—well the law of gravity is not a good example because everyone accepts that.'

He backtracks. 'But I mean science moves from—it starts with a consensus, congeals and everyone agrees that such and such is the case. And then someone else comes along around says no, hang on, I don't think it's quite like that . . . I mean science essentially should be sceptical all the time,' he says.

'Well I know *you're* sceptical,' I reply. 'You've said that parts of evolution should be challenged.'

'Oh how sweet, Anna, you've been reading Wikipedia,' says Booker. 'How much do you actually know about what I've written about evolution?'

'I've read some of your columns,' I say. 'Actually I was reading this morning about the asbestos issue—but I don't know if you still stand by that.'

'Well I suspect you don't know a huge amount about the background of these things, because I have *enemies*,' says Booker. 'I have enemies in the asbestos world, because there is a huge racket involved in trying to confuse one kind of asbestos, white asbestos or chrysotile, with two very dangerous kinds of asbestos . . . Now I have read the science, Anna, you haven't.'

'But hasn't the UK Health and Safety Authority said that your views on white asbestos are misleading?' I ask. I know for a fact that the Health and Safety Executive (HSE) has issued statements correcting Booker. Here's an example from their director general in 2002: 'Christopher Booker's articles on the dangers of white asbestos are misinformed and do little to increase public understanding of a very important occupational health issue.' Six years later, the HSE still had to issue corrections, stating 'HSE does not exaggerate the risks of white asbestos cement fibres as claimed by Christopher Booker. The article was substantially misleading . . . Finally, HSE in no way promotes the interests of the asbestos removal industry and it is absurd to suggest otherwise.'

But mentioning the repeated rebuttals from the Health and Safety Executive doesn't concern Booker. 'What they say as politicians or as officials and what their scientific papers say are two totally different things.'

Booker then goes on to defend his views on evolution. 'I have read, unlike very few people these days, Darwin's *The Origin of Species*. And I know that, I know all about the history, well, sorry, I don't know *all* about the history of evolution but I have lived with evolution, the story of evolution, all my life,' he says. 'The question is, how did it come about?' he asks. 'Darwin had one very simple explanation for it. If you read his book you will find four times he raises . . . objections to his own theory.' Booker thinks these objections were never satisfactorily answered. 'Have a look at it and see whether you think he's actually given an honest and plausible answer to his own objections because the fact is it is more complicated,' he says.

George Monbiot, a columnist for the *Guardian*, has looked at Booker's claims on evolution. His verdict on Darwin's so-called unanswered objections is that Booker appears 'unaware that all these questions—the tiredest old creationist canards—have been answered many times over by evolutionary biologists'.

But perhaps Booker raises a good point: I *haven't* read Darwin's *The Origin of Species* for myself. Despite this, I still accept the Darwinian explanation of evolution. Why is this? 'We're all incredibly lazy about the way we hold our opinions,' says Booker. 'We don't actually question them enough. I think it's a great duty on all of us to say, "Why do I believe what I do?" about almost anything.'

Booker is right. I have been lazy in not going to the original source material before I decided to accept Darwin's theory of evolution. I've never given a second thought to whether Darwin's explanation for evolution is correct or not, until today. But I'm not concerned about this: as Naomi Oreskes pointed out when we met her, all social relations are trust relations. 'We can make our own lunches, tie our own shoelaces, but most of us cannot do our own science,' she writes. Not every school student learning the theory of evolution in school *needs* to read Darwin, because other scientists have, and over the years the institution of science as a whole has accepted it to the point where (despite the best efforts of Creationists) it's come to be accepted fact.

'You bring up the issue of trust,' I say to Booker.

'Yes,' he says.

'And I guess part of this whole program is, well, who do we trust on the issue of science?' I ask. 'Do we trust the national academies of science, do we trust NASA, do we trust the Bureau of Meteorology— all of whom say climate change is happening and that it's caused by humans. Or do we trust people like yourself and the small number of scientists who agree with you?'

'Please, Anna,' says Booker. 'It's funny how often I get involved in this argument with people from your side and they say, oh, on the one side there are all the serious scientists in the world with all the great institutions, and then are we to believe them or are we to believe you?' He pauses and straightens his posture. 'No, I come to my views not because I personally am in any way an expert on any aspect of this very, very complicated subject. What I am quite good at . . . is listening to all sides, saying who do I think are telling a story which really connects with the observed data.'

Booker thinks he hears a more compelling story from the climate sceptics. He also seems to think there are a greater number of serious scientists who reject climate science than I'm aware of. 'It's not as it's often presented to be,' he says mysteriously.

I ask why, if this is the case, all the studies I've seen show that climate scientists almost all agree that climate change is happening and caused by humans? Booker likes this question. It gives him a chance to return to criticising the media.

'This question of . . . the absence of a debate in Great Britain,' he says, 'is a function of what some—I won't call it *conspiracy*—but a media *consensus* that the debate is over.' He's concerned that the truth is being withheld from the public. 'So how do people get both sides of the debate here? They don't.'

I tell him the situation in Australia is very different. It's no secret that a large section of our media has been at war with climate scientists for a long time. Before Australia's former chief scientist Penny Sackett, resigned in March 2011, she told a Senate committee that her greatest concern was that the conclusions of climate scientists were not being communicated effectively to the public. When the new chief scientist, Professor Ian Chubb, was asked about the media coverage of climate science, he described it in typical understated Aussie male fashion as 'very ordinary'.

Sackett and Chubb are right to be concerned. Most Australians are constantly exposed to a barrage of anti-environment, anti-science rants on talkback radio, in the tabloid press, and especially in the pages of our only national newspaper for a general audience, *The Australian*.

A content analysis of news coverage on climate change in Australia in the first half of 2011 found that only twelve per cent of climate change stories focused on the science of climate change. The rest focused on policy issues, mostly the debate over the price on carbon pollution. The study, conducted by the University of Technology, Sydney, found that most Australians 'receive very little reportage of the peer-reviewed work of Australian or international climate scientists'. The study also found that coverage appeared 'scarce and shallow given the fact that there were major updates on climate science . . . during this time'.

The researchers found that 'climate sceptics continue to get major favourable coverage in Australian media' and stated that 'even allowing for the "distorting lens" of news values, the coverage of climate science in Australia does not reflect the overwhelming scientific consensus that global climate change is being caused by the activities of human beings'. The report concluded that 'the presence of the views of climate sceptics in Australian news media is far greater than would be expected if conventional news practice of relying on peer-reviewed science sources was followed'.

The study also found heavy negative bias in press coverage of climate change policy, with negative coverage outweighing positive coverage by seventy-three per cent to twenty-seven per cent. 'Some of Australia's leading newspapers have been so negative in their reporting of the Gillard government's carbon policy it's fair to say they've campaigned against it rather than covered it,' states the report.

I try to explain this to Booker. 'The reporting in Australia is very different. I mean, we have so many climate sceptics in the media, particularly in our major newspaper.'

'Thank God for that,' says Nick.

'The real debate,' says Booker, 'is the one that's going on in the internet, which is the true forum of scientific discussion in our time.'

'Not peer-reviewed literature?' I ask. This seems like an extreme claim.

'Some peer-reviewed literature,' Booker concedes. 'No, I mean really the interesting thing is the arguments. The peer-reviewed process has been so corrupted, it always has been, funnily enough ... peer review is not something I attach huge importance to.'

Peer review is the process of subjecting scientific work to the scrutiny of other experts in the field before it is published. Contrary to Booker's opinion, peer review is one of the pillars of good science. Being assessed by fellow experts is how poor research is caught before being published. It upholds scientific standards and prevents unwarranted claims being given credibility. Serious academic journals simply won't publish work that hasn't passed the peer-review process.

'Anna, the story's moving on,' Booker says. 'You came into it a few years back at a time when it seemed absolutely the science was settled and no one was ever going to disagree with it.'

'When did I come into it?' I ask. He seems to think climate change is the latest cool young person craze, like hipster fashion.

'Well I assume, because I know you're very young, but I assume you've been at it for some years,' says Booker.

'I learned about climate change in primary school,' I tell him. 'I learned about the greenhouse effect in Year 5. Then in high school, we learned about the beginnings of what was happening to the climate because of human [emissions of] fossil fuels. But I've been involved in this issue for about fourteen years now. It's not, as you say, some kind of fad that I just learned about in the last couple of years. I'm deeply concerned about this.'

Our discussion is coming to a close.

'What if you're wrong?' I ask Booker.

'What if I'm wrong?' he repeats back to me.

'Yeah, on climate change,' I say.

'What if *you're* wrong?' he asks me. 'I mean we can ask, you know. The trouble with this debate is it's never a debate, because the two sides keep on talking past each other all the time.'

I'm very happy to answer him directly. I agree that there's no point talking past each other. 'Well, if *I'm* wrong and your worst case scenarios of the economics are right,' I say, 'then the world will be economically damaged and eventually will recover. If *you're* wrong and the worst case scenarios of climate science are right humans might not be able to survive. I know which one I'd prefer.'

'I'd sign up for suicide, death with dignity,' says Booker, smiling. 'No, no, I don't actually agree. I think within the next five years— you're still very young and about to get married and you've got your life ahead of you—and within five or ten years the way things are moving, I predict that the whole global warming scare, panic, alarm, will have been begun to fade into the past.'

Could he get any more patronising?

'Look, I *wish* climate change wasn't real,' I say. 'I know a climate scientist who says she wakes up every day wishing that she was wrong. It's the only branch of science that does that. But unfortunately the evidence is pretty overwhelming, so I guess we'll disagree.'

'We do,' says Booker. 'Delight to meet you, Anna.'

I mean it wholeheartedly when I tell him it was good to meet him, too. I have to forgive him for being condescending. He is, after all, old enough to be my grandfather.

And to be fair, Booker *has* given me something to think about with his point about people holding their opinions lazily. Whether it's shock jock Alan Jones' listeners unquestioningly accepting what he says on the radio, or students accepting what their science textbooks say about climate change, few people interrogate their own views and why they hold them.

I'm grateful for the opportunity this journey has given me to dive into the details of climate science. I've done what Booker would have wanted me to do—read scientific papers and IPCC reports, looked at multiple strands of evidence for climate change and struggled through dense science written by NASA climatologists. I've also looked at the sceptics' arguments, learned them inside out, evaluated them and compared them with what mainstream science says.

At the start of this journey, I accepted climate change science largely because of the weight of authority behind it. Now, I accept it because I've looked in excruciating detail at the arguments put forward by scientists, and those put forward by sceptics. I've judged for myself which arguments make sense and are based on logic and evidence.

So I agree with Booker that more people should form their judgments after looking at the evidence. But, in *light* of this evidence, I have to disagree with him on almost every other point he's raised.

Chapter Eighteen

OF COSMIC RAYS AND COMPASSION

THE NEXT MORNING BRINGS A 5.15 AM START AND A NEW COUNTRY:
Switzerland. Today's destination will be the last of Nick's picks for
the journey. He's chosen to take me to the Geneva headquarters of
CERN, the European Organisation for Nuclear Research. We're
going to take a look at one of Nick's alternative theories about what
is causing climate change, given he doesn't support the idea that
carbon pollution is the major culprit.

So far Nick has put forward the view that climate change could be
caused solely by variations in the sun. This was the theory proposed
by Joanne Codling and David Evans, the first of Nick's choices for
interviewees. Nick has also stated that perhaps climate change is
not 'caused' by anything at all, that it's merely natural variability
in temperatures. That's what Richard Lindzen suggested. Today,
Nick is proposing the view that climate change might be caused by
cosmic rays.

When Max told me that Nick wanted to take me to CERN to
discuss cosmic rays, I thought he was joking. But he's not. Apparently
cosmic rays from outer space are a potential culprit for a small

percentage of the climate change affecting us here on earth. So here we are at the CERN campus just outside of Geneva, about to meet a scientist who specialises in cosmic rays.

Kate goes into the lobby to clear our group of six through security. We're each issued a small blue press pass to carry with us at all times. We're in Switzerland, in a nuclear research centre, so everything is very organised. We are directed to the cafeteria to wait for the scientist we'll be spending the day with. I'm expecting some kind of little canteen selling sandwiches and fruit, but I walk through the doors and am dumbfounded. We've just stepped inside a modern, state-of-the art cafeteria set up to cater for thousands of hungry scientists. Tables are laden with sushi, roast beef, pizzas and pastas, salads, cold meats, breads and desserts. I remember enough French from high school to order some macaroni and cheese, then look around for Kate. She's outside, defending a table from the hordes of scientists searching for a seat.

As I eat, hundreds of scientists—probably some of the smartest people in the world—stream out of the surrounding buildings. Most of them are deep in conversation. Some carry large satchels overflowing with papers while others lug around laptops. The ones who can't find a seat just eat standing up. I've never seen so many people proudly wearing sandals and socks in the one place. It's like nerd heaven.

CERN is a physicists' Mecca. Not content with just observing nature, CERN is the place that pries free its secrets. Atom by atom, the scientists here use the world's largest particle physics laboratory to study the electrons, neutrons and protons that make up all matter on earth. The organisation started its research in 1954 and now exists due to a major cooperative effort between twenty European countries. Its main function is to provide the particle accelerators and other infrastructure needed for high-energy physics research. It fosters collaboration between scientists at 608 universities and research facilities all over Europe.

I remembered CERN making headlines across the world in 2008 for a controversial experiment. Scientists were planning to use the Large Hadron Collider here to smash atoms together in the hope of replicating conditions that existed a fraction of a second after the Big Bang. This caused concern among a small number of scientists, who were scared the experiment could create a mini black hole with the potential to tear the earth apart. A group of them even sought a legal injunction in the European Court of Human Rights to try to stop the experiment. A German expert in chaos theory claimed that 'CERN itself has admitted that mini black holes could be created when the particles collide . . . it is quite plausible that these little black holes survive and will grow exponentially and eat the planet from the inside.' The experiment went ahead and luckily, as you may have noticed, no black holes destroyed the planet.

The scientist Nick has arranged for us to meet today is an experimental particle physicist named Jasper Kirkby. Jasper comes over to our table outside the cafeteria and introduces himself. He's of medium build with a British accent, pale skin with short brown hair, the inevitable sandals and socks, and glasses. Jasper proudly tells me he has two daughters around my age, both of whom have followed their father's footsteps into science. Jasper's gentle manner and enthusiasm for his work reminds me of my own dad, who has just retired from teaching library practice at TAFE.

Jasper's research project is called CLOUD: Cosmics Leaving OUtdoor Droplets. We walk with Jasper to a large building that looks like an aeroplane hangar. There's a big yellow and black radiation warning sign out front, next to a white tank of hydrogen. Two blokes in blue hard hats smoke cigarettes nearby. We go inside and Jasper shows us his laboratory. It's a large stainless steel container—around three times as tall as a human—in which scientists have the ability to simulate the atmosphere. This is the container in which Jasper's experiments occur. It is completely free of any contamination, so that the experiment can't be corrupted. 'It's the cleanest chamber in the world,' says Jasper proudly.

'So if, for example, a fly got in there would it contaminate the experiment?' I ask.

'Oh, that would be terrible, yes,' says Jasper. 'It would be terrible for the fly as well.'

Into this super-clean environment, Jasper beams artificial cosmic rays created by powerful particle accelerators. The idea is to see whether these rays have the power to seed aerosol particles. If they do, and if these aerosol particles can get big enough, they have the potential to create clouds. Clouds are significant to climate change discussions because they block sunlight reaching the surface of the ocean and the earth. Fewer clouds mean more sunlight. According to Danish scientist Henrik Svensmark, fewer clouds due to reductions in cosmic rays might be the reason the earth is heating up (rather than more greenhouse gases). Jasper's experiment is set up to test this hypothesis.

Before we go any further, I have to ask Jasper to explain what cosmic rays actually are. 'Cosmic rays are produced by supernovas in the Milky Way galaxy,' he says. I look at him blankly, so he elaborates. '[Supernovas are] exploding stars, and these stars throw out a huge amount of particles and after a few million years or so, if they're pointed in just the right direction, they arrive on the earth's atmosphere and then they rain down and create this continuous level of radiation through the atmosphere.' I think I understand. Cosmic rays are a kind of radiation caused by exploding stars millions of years ago that travel to earth at the speed of light.

The question Jasper is investigating is, when cosmic rays arrive in the earth's atmosphere, what do they do? The reason Nick has taken me here is to imply that they might be responsible for causing climate change. Any uncertainty around the science gives policy makers a reason to delay action. Experiments like CLOUD are valuable. However, they can and are being used by climate sceptics to argue that we just don't know what's causing climate change, and therefore there's no point reducing carbon pollution.

'What we want to do at the end of the experiment is really have a definitive answer—yes or no—cosmic rays do or don't affect the climate,' says Jasper.

'You're confident you can get an answer?' asks Nick.

'I'm absolutely confident CLOUD can do that,' Jasper replies.

Nick and I ask a few more questions to ascertain what the experiment is looking at, and what it isn't.

'At the moment we say nothing about the effect of cosmic rays on climate,' says Jasper. 'But over the coming months and years, hopefully a few years, we'll answer it very definitively.'

The reason cosmic rays could contribute to cloud cover, so Svensmark's argument goes, is that they might be able to create aerosols. Aerosols are just another word for 'fine particles'. They mask temperature increases due to climate change by reflecting heat into the atmosphere and creating a temporary cooling effect.

When they get big enough, aerosols can also create clouds, which block sunlight from reaching the earth.

So if there has been a reduction in cosmic rays, there will be fewer aerosols and therefore fewer clouds (if Jasper's experiment proves cosmic rays *do* form aerosols big enough to turn into clouds). According to the sceptics, this could mean that warmer surface temperatures are *not* due to greenhouse gases trapping heat, but simply due to fewer clouds letting more sunlight in, because of a reduction in cosmic rays.

But the experiment isn't yet at the point of looking at clouds. 'We're at a very early stage,' says Jasper. 'We've only looked at the effect of the cosmic rays on the formation of very, very small particles.' The results of his recent experiment suggest that cosmic rays do indeed seed tiny aerosols. However, Jasper doesn't yet know whether these aerosols are big enough to go on to form clouds.

'So what actually have you established?' asks Nick. 'That there's a mechanism by which cosmic rays are seeding?'

'Yes,' replies Jasper. 'But so far we've only looked at . . . the formation of particles up to a few nanometres size and they have to grow much bigger before they're capable of seeding cloud droplets.' In fact, studies have shown that 'freshly nucleated particles' (which is what Jasper found cosmic rays can create) must grow by approximately a factor of 100,000 in mass before they can effectively scatter solar radiation or be activated into a cloud droplet.

'What's *not* established is whether cosmic rays do indeed affect clouds,' says Jasper. 'And that's what we want to settle with this experiment.'

Here's the argument posed by sceptics: *if* cosmic rays cause clouds, and *if* there have been fewer cosmic rays since the Industrial

Revolution due to a stronger solar magnetic field, then there will have been fewer clouds and therefore more sunlight hitting the earth. More sun means warmer surface temperatures. Therefore they argue that cosmic rays, rather than greenhouse gases, determine how warm or cold the earth gets.

There are a lot of 'ifs' in this theory. And while Jasper is testing one of the three steps required for cosmic rays to form clouds, other scientists have measured the strength of the solar magnetic field and the number of cosmic rays reaching earth over the past sixty or so years.

Unfortunately for climate sceptics, these patterns have not behaved in such a way that would attribute climate change to less cloud cover due to fewer cosmic rays. A peer-reviewed paper by Mike Lockwood in 2008 concluded that cosmic ray patterns 'have been in the opposite direction to that required to explain the observed rise in mean temperatures'.

Even if Jasper's experiment proves that cosmic rays do cause clouds, they can't be the primary cause of global warming. I email the eminent Australian atmospheric scientist Professor David Karoly to see if he can help me explain this in simple language. Are cosmic rays a major factor affecting global temperatures, I ask? And if not, why not?

'There is a large eleven-year cycle in the amount of cosmic rays received at earth,' he responds. 'The solar sunspot eleven-year cycle modulates the solar magnetic field and the earth's magnetic field, affecting the number of cosmic rays received at the earth over time.' He tells me that the amount and intensity of cosmic rays in the earth's atmosphere has been measured monthly since 1953, and this eleven-year cycle is clearly shown.

David Karoly explains the implications of this for the sceptics' cosmic rays argument. 'The large eleven-year cycle in cosmic ray numbers leads to a peak to peak variation in cosmic rays of about 10% to 20% over each eleven-year period at all the long term observing sites,' he writes. 'If there was a long term influence [of cosmic rays on global average temperatures], then it should be apparent when cosmic rays vary most, over the eleven-year cycle.' What he is saying is that if cosmic rays were driving the temperature (rather than carbon pollution) we would see temperatures change in sync with this eleven-year cycle. But he tells me this temperature change 'is *not* found globally or at high latitudes, where the cosmic ray variations are largest'.

This doesn't mean that cosmic rays don't play some role in the natural variability that's constantly in the background of human-induced climate change. And it doesn't take away from Jasper's important work showing that cosmic rays play a role in seeding aerosols that have the potential to make clouds. The more we understand the climate system, the better, which is why I'm genuinely receptive to Jasper's work. I'm just not happy at the way sceptics have used his results to argue it proves something that it does not.

Jasper doesn't challenge the science that shows human emissions of fossil fuels are responsible for most of the climate change the world has experienced post Industrial Revolution. 'We certainly know that carbon dioxide is affecting the warming of the planet,' he says. 'What we don't know is how much it's affecting [it].'

As I noted earlier, the Intergovernmental Panel on Climate Change is clear that greenhouse gas emissions are responsible for *most* of the observed warming. They haven't said *all*. So what Jasper is looking at is the contribution from natural causes. 'We really don't know what the natural contribution is,' says Jasper.

Jasper says his experiment 'could change very much how we understand the natural contribution to climate change—which will *add* to whatever else is going on'. He clarifies: 'It won't remove the other sources. It will be a new source that's presently unaccounted [for].' In other words, Jasper is looking for a new climate forcing agent in addition to the ones we already know about, such as human emissions of carbon pollution.

'So if we do find a link between cosmic rays and clouds . . . this *adds* to the forcing agents that [are] causing climate change,' Jasper says. 'It doesn't *remove* the other forcing agents. It means that we [might] have a new one, a natural one that may or may not be very significant.'

I tell Jasper that this is not how his results are being interpreted by climate sceptics. 'A lot of the climate sceptics are saying that your experiment proves that humans aren't creating climate change, that cosmic rays are. Is that right?' I ask.

'No, that's an extreme point of view,' he says.

'No one's saying that,' says Nick. I guess he hasn't read the sceptic blogs.

'We've identified one important climate forcing agent for the twentieth century and that's the change of carbon dioxide, and it will certainly be having an effect on the climate,' says Jasper. 'What we don't understand is what the natural contribution is.'

'What happens, though, when research like this is misinterpreted?' I ask. 'I have read some blogs saying this [experiment] proves that climate change is not due to humans at all.'

'This goes on both sides,' says Nick. 'You know, there's always going to be arguments both sides.'

'It's up to society,' says Jasper. 'Science provides no answers about what to do with that knowledge. We can only provide knowledge and then society has to decide how to handle it. The classic example here is nuclear power. Nuclear scientists at the end of the Second World War developed the atomic bomb. That knowledge can be used for a horrible thing like a bomb or it can be used for . . . unlimited power for humanity and it's up to society to decide how to use that knowledge.'

'I guess I'm just wondering how you feel about it,' I ask, referring to the way his experiment has been used by climate sceptics.

'That's the public debate, that's the public debate,' says Nick. I think he's trying to communicate that it's perfectly legitimate for Jasper's results to be used and interpreted by whoever wants to do so.

'What happens in the blogosphere is, somebody starts a rumour here and it goes viral and very quickly everybody repeats a rumour,' says Jasper. 'It's a Chinese whisper. You know, that's not science. That's something in the blogosphere. But our job as scientists is to do the best science . . . We can't control what's said of our results.'

Nick changes the subject to another of the blogosphere 'rumours' surrounding Jasper's work. 'There are reports that the director general here was very strict with you in what you could say about all this. Is that correct in terms of its implications?' asks Nick. I've seen it said on climate sceptic blogs that Jasper's results were censored by CERN's director general to make them 'fit' the climate change orthodoxy. For example, Anthony Watt's blog *Watts Up With That* says: 'The Director General of CERN stirred controversy . . . by saying that the CLOUD team's report should be politically correct about climate change.'

'No,' says Jasper. 'I know where that was started and it was a complete misinterpretation. The director general has never controlled [or] limited any scientific information from the lab.'

Nick says he's pleased to hear that.

•

In the car after the meeting, stopping and starting every few minutes in peak hour traffic, I think about Jasper's views on the role of science. Jasper thinks that scientists should keep out of public debates around the implications of their research. 'It's not up to scientists to advocate action,' he'd said. It is a perfectly legitimate view. But in light of what's at stake, this seems naive to me. Scientists have a moral compass, just like the rest of us. Is it really OK for them to just sit back and watch the debate unfold, when they know what they do about what's happening to our planet?

For many scientists these days, their sense of stewardship kicks in and prompts them to speak out about the implications of their research. One comment by a professor of physical sciences on a blog post sums it up: 'We just need to keep banging back at them when they lie and distort [our research] . . . Scientists are in the ring now even though it is not part of our training and we feel there are better things we could be doing.'

The eminent climatologist Steven Schneider echoed this sentiment. Before he passed away in 2010 he left a message on his climate science website with reflections on his decades of work in the field. He wrote: 'Just because we scientists have PhDs, we should not hang up our citizenship at the door of a public meeting.'

Chapter Nineteen

THAT SINKING FEELING

OUR NEXT MEETING IS WITH IRENE KHAN, A BANGLADESHI WOMAN who recently retired from her role as secretary general of Amnesty International. I've chosen her in an attempt to open Nick's eyes to the suffering climate change is *already* causing for the world's poor. I'm aiming to help Nick grasp that for many people in the world, climate change has nothing to do with an ideological discussion about government intervention in the free market. Instead, it's about life or death.

From what I've seen of her on television, Irene is strong and articulate. She is a trustee of Mary Robinson's Climate Justice Foundation, an organisation set up by the former president of Ireland to focus on the way climate change is undermining human rights and unfairly impacting vulnerable communities. Given that the people worst affected by the impacts of climate change are from developing countries like Bangladesh, I'd fought hard to get the production team to agree for us to meet Irene.

Max and Simon Nasht initially hadn't been interested in using her as a spokesperson, arguing that I already had enough people and that the show was about the science of climate change, not the impacts on people in poverty. Then I'd learned that Nick had chosen

Bjørn Lomborg. I insisted that I have the opportunity to find a spokesperson who could rebut the argument that Bjørn often poses, that it's more important to solve global poverty than it is climate change. They were also counting Nick's two spokespeople in Perth, Joanne Codling and David Evans, as one person. I pointed out that that was not strictly true. I wouldn't back down, and the production team had to reluctantly agree.

By lucky coincidence, Irene happens to be in Geneva at the same time as we are. She's just flown in today from London. So Nick and I check into our hotel and wait for her at the bar. Nick downs three beers and I have a glass of water. I'm already exhausted from our 5.15 am start and I suspect a beer would send me straight to sleep.

Irene is immediately recognisable when she walks in. I've seen her on the news plenty of times in her former role leading the world's largest and most influential human rights organisation. A diminutive, olive-skinned woman with curly dark hair, she walks with grace and poise. She sits between me and Nick, in an elegant camel-coloured top and a blue blazer, and begins by talking about the impacts of climate change on the poor.

'For me, I see it through the life I lead and the people I meet,' says Irene. 'I've visited the coastal areas in Bangladesh and I hear people talk anecdotally—not scientifically but anecdotally—about how their lives have changed because of more frequent flooding, because of erratic rainfall, because of drought, because of the coastline shifting.'

She talks about climate change's impact on the poorest of the poor, noting that 'the saddest part of it is that they just don't have the capacity to cope with the change that's taking place'.

I raise the argument often made by climate sceptics, that the world faces a choice between solving climate change and solving poverty. Two of the people Nick has taken me to meet are known for making this argument. As I've already mentioned, Bjørn Lomborg spent years arguing that funds would be better spent on malaria, poverty and HIV/AIDS than climate change. Joanne Codling also argues that acting on climate change is bad for people in poverty. In her *Skeptics Handbook* she argues:

> *If we make it harder or more expensive for people in Africa to use their coal, it means they keep inhaling smoke from wood fires; babies get lung disease; forests are razed for fuel. Meanwhile electric trucks cost more to run, and that makes fresh food more expensive; desperate people eat more monkeys—wiping out another species; children die from eating meat that's gone off . . .*

Irene shakes her head at this line of reasoning. 'These are not linear processes,' she says. 'I have a major problem with those who think that global challenges are [separated] in silos and that you deal with them silo by silo. The real challenge today is that all these issues are interconnected.'

Irene talks about the problems faced by countries like Bangladesh. She mentions infrastructure, health and food prices. 'What climate change is doing is aggravating the problems they [developing countries] already have,' she says. 'You have to deal with the problems in an integrated way. Because climate change is happening here and now. It's undermining whatever efforts are being made to address poverty.'

Nick asks Irene about the flooding that's affecting Bangladesh. 'It's happening already,' says Irene. 'One [cause] is the melting glaciers that bring more waters down the rivers, and the other is the rising sea levels.' Bangladesh is stuck smack-bang in the middle of two major climate change impacts.

Nick makes the point that perhaps, then, money should be spent on adaptation. Irene agrees that this must happen, but adds, 'But I don't think that means we don't do anything else. We need to do adaptation *and* mitigation.'

'All those things are still not going to be enough,' she continues. 'It's like putting a bandage on. At the same time you need bigger solutions. What's happening in Bangladesh is not because of something Bangladesh has done.' It's a hard truth for a nation to face. No matter what they do in their country they cannot protect themselves from the impacts of a problem that is truly global.

Irene is warm and personable. She seems cautiously optimistic. 'I trust young people, the next generation,' she says. Despite this, her dark eyes contain flashes of sadness. Perhaps that's because the situation her country is facing is just so staggeringly, blatantly unfair. On average, a Bangladeshi emits only 0.36 tonnes of CO_2 per person. Americans emit almost 18 tonnes, and Australians a staggering 19.6 tonnes. China, by contrast, emits 5.8 tonnes per person and the average Indian only 1.38 tonnes. Yet the worst impacts of climate change are felt in countries like Bangladesh. That 1 metre sea level rise by the end of the century that Admiral Titley talked about? For Bangladesh, this wipes out almost twenty-one per cent of its land, exposing almost 15 million people to inundation by the rising tide. India would lose just 0.4 per cent of its land, but this equates to 7 million people—over a quarter the population of Australia.

Nick talks about India and China and how they're going to emit large amounts of carbon pollution as they develop. Irene makes the point that countries like India and China don't have to develop using the same energy technology that the West used in the 1800s and 1900s. 'We don't have to make the same mistakes as our grandfathers,' she says. By now, both technology and our understanding of the environmental risks of using coal for power have evolved.

'I don't think environmentalists are saying don't develop,' says Irene. 'Environmentalists are saying develop in a different way. Develop in a way that is sustainable. That protects the planet. Because if the planet isn't there, then nothing else will matter.' She talks about alternative pathways of development that 'leapfrog the dirty cycle [of fossil fuels]'.

'That is one of the conundrums in this,' says Nick. 'Australia is one of the world's biggest coal exporters and two of our biggest growing markets are India and China. They desperately want to provide electricity to their people. So what do we say?' he asks.

'What I would like to hope is that the *whole world* can find cleaner energy,' Irene answers. She talks about the responsibility of rich countries like Australia to not just reduce our emissions, but also support developing countries with clean energy technology.

'But they're just incredibly expensive,' argues Nick.

Irene acknowledges that there's a cost, but says 'it's a cost that has to be paid. India and China are not going to come on board until the West actually makes the effort.'

This leads us into the topic of climate justice. 'Climate justice is the concept that there has to be equity in addressing the issue of climate change,' explains Irene. 'Because those who have benefited from greenhouse gas emissions and those who are emitting, are not the ones who will feel the worst impact of climate change.'

'The poorest are going to be affected the worst because they have the least capacity to adapt,' says Irene. She hopes that climate change will catalyse a new global mindset. 'Climate change is forcing us to think globally, because we're talking about the planet,' she says. 'When I look back a generation to my grandparents they were bound not just nationally but by their cities and villages. We are now beginning to think beyond that.' I hope she's right.

I ask Nick to try to look at what's happening from the perspective of our neighbours in Bangladesh and the Pacific, putting ourselves in their shoes. Irene points out that empathy plays a role, but it's also in developed countries' self-interest to solve climate change.

'We're all in the same ship. And the ship is sinking,' she says to Nick. She talks about the issue of environmental refugees. 'I know Australia is very worried about migration,' she says. 'People are not going to be sitting there [in countries like Bangladesh] *waiting* to go under water. They're going to move.' Estimates of the numbers of climate change refugees, or people displaced from their homes for environmental reasons, range from 250,000 by the middle of this century to millions of people over a longer period of time. 'You already see this pressure happening in parts of Asia and Africa where large numbers of people are moving,' says Irene. 'And then you see all the resource conflicts taking place over water, over land, and food prices going up. All sorts of tensions between people.'

Irene finishes by linking her current work on climate change to her previous role leading Amnesty International. She shares a list of basic human rights that climate change is already affecting. 'The right to life, the right to food, the right to health, the right to water—all that is going to be affected by it,' she says.

After Irene leaves, I think about her work leading a movement of hundreds of thousands of people. In the space of two generations, human rights have gone from a marginal to a mainstream concern. It gives me hope that we can build the same base of support for action on climate change.

In the past decade, I've seen increasingly stronger links and cooperation between human rights, aid and development, indigenous justice and women's rights organisations. 'Climate change and global poverty are challenges that must be addressed together,' states World Vision's website. 'Climate change is an even greater threat to us than apartheid was, because as temperatures rise, millions of Africans will be deprived of water and crops,' stated the South African Archbishop Desmond Tutu. 'This will cause enormous suffering.'

It's true that the climate justice movement is growing stronger and deeper. The growth of the Australian Youth Climate Coalition is a case in point. Our local groups are just as strong in regional and rural areas as in the capital cities. Our membership is diverse and cuts across race, class and geography. We're building a new constituency who will shop, act and vote on the basis of climate change.

Blocking the path of this swelling movement for change stand just two things, closely linked. People with vested interests in the fossil fuel industry cling to an economic system set up to deliver them record-breaking annual profits. By their side stands the small remaining band of climate sceptics. Together, they continue to stall action on climate change by standing in the path of a new economic system based on clean energy. Their arms are linked and they shake their heads firmly. Their mantra? 'I remain to be convinced.'

I know that anger isn't the most useful emotion, but sometimes it's hard not to be angry at Nick and his exclusive club of sceptics. Their arguments don't stand up to intellectual scrutiny, yet these arguments have been holding the world back from cutting carbon pollution for decades. Nick seems be content with delaying the switch to clean energy by searching languidly for an explanation for climate change—any explanation other than carbon pollution, that is. Some of his spokespeople have said it's the sun; some have said it's just natural variability; and today he's implied that cosmic rays are to blame. It's like he'll entertain any explanation other than the sensible, scientific, rational one. He seems determined to avoid facing the blindingly obvious.

Nature and science have the answers to all of Nick's questions, of course. We've already learned a lot from the ice cores, and the changes in natural systems today are full of messages about what's happening to our planet. But nature takes time to fully reveal its secrets, and time is the one thing we no longer have on our side.

Chapter Twenty

COLLAPSING CLIFFS

WE ARRIVE AT THE UNIVERSITY OF EAST ANGLIA TO GREY SKIES. WE'VE flown from Geneva back to London, then taken a three-hour van ride to the medieval town of Norwich on the east English coastline. Nick and I haven't been told who we're meeting here, as it's another of the producers' picks. Our van pulls up outside a big concrete slab of a building. It looks like it belongs in eastern Europe, not a small English village. The sign out the front reads The Tyndall Centre for Climate Change Research.

A man with a beaming smile comes out to greet us. With his greying hair, rosy cheeks, glasses and moustache he looks like a librarian or a professor. Unsurprisingly, it turns out he's the latter. Professor Mike Hulme is the founder and former director of the Tyndall Centre, an interdisciplinary research institution concerned with the social, political and economic ramifications of climate change. Mike no longer works at Tyndall, but in 2009 he published a book called *Why We Disagree About Climate Change*.

Mike is slightly controversial in climate change circles, since he's much more conciliatory towards climate sceptics than most scientists are prepared to be. Mike argues that environmentalists should settle for an incremental and piecemeal approach to decarbonising the economy, instead of aiming for a legally binding global treaty to reduce carbon pollution. He urges climate campaigners to focus on policy steps that climate sceptics can agree to, such as energy efficiency. This is why the producers have chosen him as a 'neutral' pick.

Max directs Nick, Mike and me to a corridor in one of the newer university buildings. It's all modern architecture and floor-to-ceiling glass walls. As the sun comes out from behind the clouds, the corridor becomes blindingly glary. The three of us sit down and squint at each other while we talk.

Mike wants to know our best estimate of how many degrees we think the earth will warm by, if CO_2 emissions reach double their pre-industrial levels. 'I'm interested, listening to what the scientists are saying, what do you think the climate sensitivity is likely to be?' he asks.

I go first. 'Well, the IPCC says it's between 2 and 4.5,' I say. James Hanson from NASA thinks it's around 3—and it seems that science is tending to converge around that number of 3, so that's probably where I'd say is most likely.'

'OK,' says Mike. 'And Nick, where would you . . .'

'Well, see, I think one of the problems in all of this debate is that the public have been told that it's just CO_2, that is, you know, going to . . . boil the planet,' says Nick, before launching into a long discussion of feedback mechanisms and their existence or lack thereof.

Mike tries to get Nick to answer the question. 'So you would say . . .' he asks.

'He doesn't believe that there are feedbacks,' I explain.

'I'm saying that remains a very big uncertainty and a big unknown,' clarifies Nick.

Mike looks mildly frustrated. He probably thought he was just asking a simple question. But he should have known that if anyone knows how to avoid a question, it's a politician.

'So if I really put you on the spot and said OK, what would be your gut instinct about what the climate sensitivity is,' he asks Nick again, 'what would you estimate it to be?'

'To the doubling of CO_2, well I'd say, at most, 1 degree,' says Nick. He goes on to talk about Richard Lindzen's proposition that the climate system is 'self balancing'. 'To the extent there is any warming effect from CO_2, one of the consequences of that will be greater cloud cover and that will have a balancing effect,' says Nick. 'So other things being equal you'd expect a neutral outcome.'

Given the temperature record shows we've already had a 0.8 degree temperature rise with a thirty-eight per cent CO_2 increase, it's strange that Nick believes we'll only have a 1 degree temperature rise with a hundred per cent CO_2 increase. Nick accepted the temperature record way back in San Francisco when we met with Professor Muller. If a thirty-eight per cent increase leads to 0.8, how can a hundred per cent increase lead to only 1? But for some reason, Nick clings to Lindzen's failed theory.

As a former climate scientist, Mike knows as well as anyone that Lindzen's theory doesn't fit with the empirical measurements. Nevertheless, Mike believes in taking a different approach to communicating the need for climate action, by talking about risk rather than arguing over the science. 'I'm trying to use the CO_2 doubling as a way into thinking about risk and uncertainty,' he says. 'Because it does seem to me that actually I can quite easily work with both of you here . . . because we don't *know* what the climate sensitivity is.'

Mike goes back to the IPCC's estimate of climate sensitivity. '[The IPCC] odds are that it's a two-to-one chance that it is between 2 and 4.5 degrees,' he says. 'There's a significant possibility it could be less than 2—let's say 10 per cent. But there's also a significant possibility it could be bigger than 4.5—let's say 23 per cent.' I think about these odds, and how people watching the documentary at home will interpret them. 'The issue we've got to deal with,' says Mike, 'is how do we deal with risk and uncertainty.'

Nick likes the word uncertainty. Climate sceptics have used it for decades as a smokescreen for delaying action to cut carbon pollution. 'I was keen to take Anna to CERN to see the work that's going on there . . . on the whole cosmic ray issue,' he says to Mike. 'He [Jasper Kirkby] made the very important point that there are so many uncertainties.'

'But he also said this is in *addition* to human influences,' I say.

'Can I finish? Don't cut me off,' snaps Nick. 'The whole area of solar activity is still a big uncertain area and is not properly accounted for in IPCC work or modelling.'

Mike diplomatically tries to steer the conversation back to a more practical focus. 'But in terms of public policy . . . it is only the *human* components of climate change that we can do something about. We can't actually manipulate the sun,' he says. 'So yes, those factors are there as part of the background variability of climate, but the pressing questions are whether we intervene about the *human* dimensions—and to what extent we should intervene.'

Then Mike tries to focus on what he sees as the two ends of the climate change 'debate'. 'Would you recognise that there could be two extreme positions?' he asks. 'One would be that we can adapt to all changes, whether they're natural or human, [that] we don't have to worry about mitigation.' By mitigation, Mike means cutting carbon pollution to reduce the problem. 'The other extreme policy position,' he says, 'would be to say . . . we really have to put *all* of our investment into the mitigation part.'

This dichotomy seems meaningless to me. It's true that most climate sceptics say we don't have to worry about reducing emissions (mitigation). In their minds, if climate change turns out to be real we can just adapt to whatever happens. For example, the US Chamber of Commerce petitioned the Environmental Protection Agency, demanding it take no action on climate change on the grounds that 'populations can acclimatize to warmer climates via a range of behavioral, physiological, and technological adaptations'. When I first saw this, I had to read it twice. Humans can *adapt* our *physiology*?

So, the anti-mitigation end of Mike's spectrum does exist. But I'm not sure the anti-adaptation end does. Climate advocates generally argue for a two-pronged strategy of mitigation (to avoid future impacts) *combined* with adaptation (to deal with the impacts happening now). In fact, this is exactly what Irene and I discussed yesterday in Geneva.

'We're already seeing the impacts of climate change—particularly in Australia,' I say. All levels of Australian government are already developing strategies to adapt to the sea level rise and infrastructure damage they know climate change will cause. 'Then the question becomes how can we *avoid* what we *can't* adapt to,' I continue. The American author Bill McKibben often uses a phrase that sums up my attitude to this question perfectly. He argues that we must 'avoid what we can't manage and manage what we can't avoid'.

Mike is keen to talk further about adaptation. At his suggestion, we pile into a few cars and drive down some very English country laneways to the small hamlet of Happisburg. Despite its cheerful name, what's happened here is anything but happy. Over the past two decades the cliffs have retreated inland due to sea level rise and erosion from wind and waves. Over 50 metres of cliffs have collapsed into the sea, taking twenty-six homes with them in seventeen years. The houses that remain look precarious. Most of the homes were abandoned a few years back when the government bought them at half the market value. 'What we've got [here] is quite a dramatic landscape of eroding cliff and falling houses,' says Mike as he points out over the ocean to indicate the place the coastline used to end before being swallowed by the sea.

We're standing on one of the remaining sections of cliffs. Mike tries to use Happisburg as a case-in-point about the problems of a simple cost–benefit approach to climate change. 'Bjørn Lomborg, for example, has a particular way of approaching cost benefit analysis,' says Mike. 'And one could try to apply that here to figure out what should we do with coastline that's eroding like this. What are the costs of putting hard coastal defences in, as opposed to the benefits that accrue to householders?' But Mike explains the limits to this approach, telling us 'the people who live here are not very happy with a simple cost–benefit analysis. They value their houses and their local village environment and you can't put a price on that.'

Nick questions whether the collapsing cliffs have anything to do with climate change in the first place. 'This is a classic case of poor planning, allowing houses to be built too close to an area that's obviously subject to significant wave and wind erosion,' he says. 'So for me it's not quite the sort of example you're suggesting, of the application of cost–benefit analysis on the issue of long-term climate change.'

Some locals might agree with Nick. After all, the cliffs have been slowly eroding for centuries, with the recent sea level rise only causing it to speed up. But others don't. One resident, Malcolm Kerby, told the *Mirror* newspaper that what's happening to the Happisburg coastline 'is the front line of climate change in Britain . . . we can't deny it. It's happening. We see it with our own eyes.'

But regardless, I see what Mike's getting at. He's reminding us that simple cost–benefit analyses miss the complexities involved in dealing with human lives. 'Some people say you can't afford to deal with climate change because it's expensive,' I say, 'but how do you look into the eyes of someone from the Pacific Islands or Bangladesh or Africa . . . they will tell you we can't afford *not* to.'

In contrast, the people of Happisburg are citizens of a wealthy country. 'Villages like this actually do have the resources to do some kind of adaptation,' I point out. 'But ultimately if we're looking at . . . a world that's 4 degrees warmer or 5 degrees or 6 degrees—you can't adapt to that.'

Nick talks about applying a cost–benefit analysis to Happisburg householders. 'If they're saying I want to stay here and you're going to have to spend £10 million building their sea wall, it might be cheaper . . . to say I'll give you £1 million for your house, you go and buy another one five miles inland,' he says. 'At the end of the day it is fundamentally a function of cost.'

'It kind of comes back to what we were talking about before, you avoid what you can't manage and you manage what you can't avoid,' I say. I'm worried Nick is missing the big picture. 'We're starting to see some impacts here that can be managed by a sea wall . . . but this is a really small little example.' I keep thinking about what Irene told us yesterday, how the capital of Bangladesh is the same height above sea level as London. 'Nick, we have major cities on coastlines all around the world where we're not just talking about a couple of houses, we're talking about literally millions of people having to make these kind of decisions,' I remind him. 'We're looking at huge losses.'

Mike, standing next to me, knows all about the risks the world faces from sea level rise. A team of modellers at the Tyndall Centre did the numbers as part of the UK government's Stern Review into the economics of climate change. For a 1 metre sea level rise above today's high-water mark (which is, remember, the scenario that the US Navy is planning for) 22.5 million square kilometres of land would be lost. This would affect 145.2 million people and destroy GDP worth US$944 billion, calculated at market exchange rates.

After reading the same Tyndall research report, author Jo Chandler shared a list of the places worst affected: 'Bangladesh, Vietnam, Myanmar and much of the Netherlands are in trouble with a 1 metre rise. Alexandria, ancient city of learning, would be wiped; Hollywood, modern centre of popular culture, would also have its feet wet.' And of course, many of the Pacific islands would no longer exist outside history books.

In Australia, around eighty-five per cent of the population lives in the coastal zone. The federal government released information in 2010 showing that parts of many heavily populated Australian cities are expected to be inundated regularly by sea-level rise. Brisbane, Perth, the Gold Coast, Melbourne and Sydney are all at risk. An estimated 240,600 buildings and infrastructure valued at A$63 billion could be damaged or lost due to sea level rises by 2100.

Significant parts of my hometown, Newcastle, and the surrounding central coast, are also particularly exposed. Ironically, in October 2011 the *Newcastle Herald* reported that the Hunter's coal export infrastructure was at risk from rising sea levels. Many of the beaches where I learned to surf—beaches I'd hoped to one day bring my own children—are expected to be regularly flooded due to a combination of rising seas and high tides. The ancient coastal dune system at Stockton Bight that I'd fought so hard so protect from sand mining when I was sixteen is listed as one of the most vulnerable areas in the region.

How realistic are sea level rises this significant? Unfortunately, very. The 2007 IPCC report estimates sea level rise will be between 26 and 59 centimetres by 2100. However, NASA's James Hansen and other eminent scientists point out that this grossly underestimates sea level rise by not incorporating the contribution from accelerating ice sheet loss. The CSIRO notes that recent observations from tide gauges and satellites are tracking at the upper end of the IPCC's projections.

A slew of reports predicting greater sea level rise based on satellite data were released after the deadline for IPCC consideration. One of the most important of these was published in the *Proceedings of the National Academy of Science of the USA* in 2009 by Stefan Rahmstorf and Martin Vermeer. These more recent estimates put predicted sea level rise for a business-as-usual fossil-fuel scenario at quite alarming levels. They estimate sea level rise of between 75 centimetres and 1.9 metres by the end of the century, with an average of between 1 and 1.4 metres.

From what I've read, sea level rise remains one of the most uncertain areas of climate science. But we do know there's a lot at stake. Antarctica and Greenland together contain enough ice to raise global sea level by 70 metres.

After the three of us stand on the cliffs for ten minutes, the skies darken and the wind starts blowing up a gale. It's too blustery to keep talking, so we drive to the nearest pub to continue our chat. It's a typical small English tavern, built in the 1500s. It still has original fireplaces and, given the weather outside, I'm tempted to ask if we can light them.

•

At the pub, we continue our conversation about risk. Mike makes the point that the whole climate change debate boils down to how much risk society is willing to take. But what Mike doesn't admit is that the people making these decisions aren't the ones who are going to suffer the worst impacts. One of the big problems with climate change is the time lag between the cause and effect. The question Mike is really asking is, how much is today's generation willing to risk the futures of their children?

Mike rolls out an analogy about home security that he's obviously practised. 'My wife tends to be much more paranoid about burglars coming into the house than I am,' he says. 'We've both been exposed to exactly the same empirical evidence. We've been burgled once but that's all. We listen to the same radio programs . . . about the rate of burglaries in Norwich.' Mike's point is that just like with climate change, people can look at the same evidence but 'have quite different attitudes to what we should do about that risk'. Mike says, 'There's no point me trying to persuade my wife to change her interpretation of the evidence, just as there's no point in her trying to change my attitude to the evidence.'

I think it's a terrible analogy to talk about the risks of climate change. A robbery, while traumatic, is *nothing* compared with the projected impacts of climate change. A better comparison would be a burglar who robs you, then floods your house. And if you apply the IPCC estimates of the risk, there's only a ten per cent chance of the burglary *not* happening, compared with a twenty-three per cent chance of it being even worse than expected. I immediately sympathise with Mike's poor wife, sitting at home terrified without locks on the windows while he's working late at the university.

Maybe Nick feels sorry for Mike's wife, too. 'Well in your case, you would just do what your wife required for a happy marriage,' he jokes.

'I think it's a matter of probability,' I say. I repeat the IPCC odds, acknowledging that there is a chance that Nick's estimate of the risk from climate change is right, but this chance is less than ten per cent. Compare this with the seventy per cent chance that the 2 to 4.5 degree estimate is correct, and the twenty-three per cent chance that we'll see the disastrous outcome of more than 4.5 degrees rise in global average temperature. 'So I'd start by looking at probability,' I say. 'But then I think there's also looking at what's the worst that could happen if he's right, what's the worst that could happen if I'm right, what are the benefits of acting, and what are the benefits of not acting.'

'Even if we take the *worst case* scenario of the economics,' I continue, 'we could, by acting on climate change, spend a lot of money unnecessarily and could cause economic harm. Or we could not act—and we could destroy the basis of life on which we depend.' How can Nick fail to understand that the risks of cutting carbon pollution are dwarfed by the risks of unchecked dangerous climate change? 'Even if we're wrong and climate change was a big hoax,' I say, 'we've still got clean energy, we've still protected our oceans, we've protected our air . . . created green jobs, been more energy efficient—[things] that we would want to do anyway.' I look at Nick. 'Whereas if you're wrong and we don't act . . .'

'Look, in any public policy area you're making some sort of assessment of the likelihood or otherwise of certain events occurring,' replies Nick. 'And you've then got to do cost–benefit analysis of the different courses of action. They're never easy . . . in democracies because vested interests accrue around particular courses of action.'

Nick Minchin, powerbroker for a political party that solicits and takes donations from the tobacco industry and mining industry, is talking to me about vested interests? This should be interesting.

'Frankly, one of the public policy problems in this area is that you've got a whole bunch of vested interests now active in radical action, particularly with government intervention,' says Nick. 'So you've got environmental groups who have latched on to this, gives them a raison d'être, you've got all the bankers because they love the idea of creating a new financial derivative, trading in air, you've got all the green alternative energy companies.'

I can barely believe what I'm hearing. The most powerful industries in Australia, including mining, coal-fired electricity generation, aluminium, cement and car manufacturers, have put millions of dollars into campaigning for over a decade to block any action on climate change. The mining industry alone spent A$22 million on its advertising campaign in 2010 against the mining tax, helping to topple Kevin Rudd from his prime ministerial post in the process. And Nick is telling me that it's environmentalists, bankers and the small number of Aussie solar and wind startups who get to decide what happens in Parliament House? I'm speechless.

But while I'm trying to get some words together, the conversation moves on to Mike's main point. 'My point is that that scientific process of exploration and discovery and surprise will never produce closure,' says Mike. 'We're just going to have to get used to making decisions under conditions of uncertainty.'

I agree. 'There will always be things we don't know about the science,' I say. Science is never a hundred per cent certain—but we can't just ignore what we *do* know, and we know enough to act.'

'Again, that's a matter of opinion that we know enough to act,' says Nick. We talk about sea level rise of a metre by the end of the century. This is what Rear Admiral Titley and the US Navy are planning for. Nick calls it speculation. 'There's no hard evidence to back that up,' he says.

Mike reminds him that earlier in our discussion, Nick said that climate sensitivity was 1 or 1.2 degrees. 'Anna would say it's probably around about 3,' says Mike. 'Now that is as best as the science can deliver. It is somewhere in the range from, let's say, 1 to 6 degrees. That *is* scientific evidence. The point is that you're interpreting the *risks* associated with that evidence quite differently.' He's right. Nick and I have looked at the same data and met the same people. 'Science will never resolve your differences,' says Mike.

We delve again into why this might be the case. Why does Nick react so differently to the majority of people who have looked at the evidence on climate change?

'It does come back to values,' I offer by way of explanation.

'And your values haven't changed as a result of this journey, I guess,' says Mike.

'Exactly,' agrees Nick. 'And anyone who's mature and intelligent is going to have presumably thought through their value system and hold to it dearly . . . [it] doesn't mean you can't ever change your mind, obviously, but you do assess information and evidence based on those values.'

Mike asks us whether the journey has made Nick and me understand each others' value systems better than when we started.

'I think Anna is someone who approaches this with considerable integrity,' says Nick. 'There are some on . . . the alarmist side who have other agendas, whereas I think Anna is genuinely motivated. She believes what she reads and is acting upon that. So I respect her integrity on the issue. I don't agree with her.'

'You don't agree with her particular set of values?' asks Mike.

'It's not so much her values, [but] her assessment of the risk,' clarifies Nick.

Mike asks me whether I have understood more about Nick's values. I think about the values shared by the small group of people campaigning against climate science, and cast my mind back over the people Nick has taken me to meet. 'It has been really interesting to see that the people that we've met who are opposed to taking action on climate change have pretty much all said that they are afraid of government intervention in their lives, and particularly in the economy,' I say.

'People have different life experiences,' I say. 'When Nick grew up, communism was this massive fear . . . that the world would be ruled by some kind of authoritarian communist regime.' I think about my childhood, blissfully unaware of any kind of global threat. I hadn't even heard of communism until Year 10 modern history. 'We didn't grow up with that fear,' I say. 'So the thought of governments intervening in a market to try and solve climate change—it's just not as scary. It actually seems a reasonable thing to do.'

Nick frowns. 'Yeah, see, I don't like being typecast like that. I don't think that is productive,' he says. 'This is the Naomi Oreskes' view of the world that those of us who are more conservative on this issue are sort of still fighting the Cold War or something . . . I mean I'd like to think . . . both of us come to this with [a] reasonably clean slate.'

But then he qualifies this. 'Yes, I'm someone who is wary of government intervention in the economy because it normally stuffs it up,' he says. 'So I am wary of, you know, grand governmental solutions to something like climate, or the proposition that the governments of the world can unite and stop the climate changing. To me as a philosophically conservative person, that's just nonsense.'

By this admission, Nick should be arguing strenuously against Tony Abbott's 'direct action' plan of government intervention to supposedly reduce carbon pollution by five per cent by 2020. Nick is very clear about his values and the fact that small government is central to them.

'And that's going to have an influence on how you look [at] and interpret the scientific evidence,' says Mike. He points out that our belief systems 'act as filters when we look at evidence'.

To my surprise, Nick admits this is true for him. 'Well, but see this is fundamental to public policy,' he says. 'I always say to incoming new MPs or candidates, the only way you can succeed in politics is actually having thought through a set of convictions upon which you can judge the innumerate number of public policy issues that will come before you.'

I agree wholeheartedly that you need a set of values and principles which guide your decisions. The problem comes when these values blind you to reality. This is something I've thought a lot about in my work with the Australian Youth Climate Coalition. I'm just as susceptible as anyone else to letting my assumptions and past experiences blind me to reality.

In 2010 I took part in a nine-month long course on adaptive leadership run by Australia's oldest charity, the Benevolent Society. At the heart of the course was a concept called adaptive leadership. Adaptive problems are those that demand changes in systems, behaviour and values. This concept is very relevant to climate change because it's not the technology the world lacks to solve climate change, but the political and social will.

One of the key concepts the program challenged us to grapple with was the concept of working with neutrality when exercising leadership. I challenged myself to be aware of what assumptions, values and past experiences I bring to situations. With this greater self-awareness, I then worked on holding them lightly enough to enable me to remove them from the equation for a while. Doing this let me see problems from perspectives other than my own. I was able to understand things I would have otherwise missed. For example, I applied the work I'd done with the Benevolent Society to help me understand Nick's values of individual self-reliance, market freedom and economic survival of the fittest. This helped me forge some common ground with him on the issue of ending fossil fuel subsidies in Australia and globally.

One of the course trainers, Geoff Aigner, explains the concept of working with neutrality in his book, *Leadership Beyond Good Intentions*. 'Our passions are linked inextricably to who we are—our identity,' he writes. 'It is never just about an issue—it's also about us.' This has been obvious through the course of this journey. It is true for both Nick and me. 'When so much is at stake it's hard to get perspective: we think "right" is on our side,' continues Aigner.

'And losing or giving up is not an option . . . so there is a pride here at play which is less than useful.'

This is something I've struggled with throughout the past three weeks. My neutrality skills have been significantly tested. So far at least, I've managed to keep from breaking down or losing my temper, both on camera and off. I've tried to put my own baggage to one side and see things from Nick's perspective. This is why I chose spokespeople who would speak to his values and who wouldn't scare him off from the word 'go'. It's meant I've been able to have more authentic conversations with him, human to human.

I've tried to do this because the conversation we are having is not new. The 'debate' over climate science has been repeated and regurgitated through the blogosphere, on radio, in newspapers and around the dinner tables of Australia until most people are understandably sick of hearing it. The factions are well-defined and the conversation predictable. Each side knows what the other will say before anyone even opens their mouths. In this context, neutrality is crucial. Aigner writes:

> The problem arises when we wish to make change, not just negotiate
> – particularly change that is lasting . . . So the convictions we prize
> and hold dear can be limiting: they are part of the opposing and equal
> forces maintaining the equilibrium and blocking systems learning.
>
> 'The wisdom of children' is a phrase I have heard when a child
> is able to see through the situation to what is really happening. This
> wisdom comes from neutrality; children have few alliances, baggage
> or attachments. Their identity is still forming. They can be awake,
> fresh and curious. Neutrality does not mean forced detachment or
> an absence of emotion. Nor is it a defence against what is going on

around us. It is energetic, alive and perceptive rather than overheated or frozen. Neutrality may not be for everyone. But if our goal is progress, we can't do without it.

Now, I'm not saying I've yet mastered the art of working with neutrality. As a young person alive today, the threat of imminent global ecological and economic collapse is obviously very personal to me. And as someone who wants a family one day, climate change will directly affect not just my life but also the lives of my children.

But I'm trying. I know that to make lasting cuts to carbon pollution, I have to work across various factions. And I can distinguish between the ideological warriors doing everything they can to oppose action on climate change, and the large number of Australians who are just a bit suspicious or unsure about the science. It's the latter group I'll work with whenever I have the chance. They don't have fossil fuel money stuffed in their back pockets. They aren't ideologically motivated, either. They just want clear information on the science and the solutions—and they deserve answers to their questions.

Mike asks whether Nick and I think democracy can resolve the political and economic tensions around acting on climate change. 'What's the alternative?' I ask. 'We don't have a workable one. We've got to work together to tackle this problem and either we'll fix it or we won't. Time will tell.'

'But that would mean that you would be willing to engage with people like Nick,' says Mike.

'Of course,' I say. 'This is what this whole project has been about.'

Nick isn't evil and I have no objection to engaging with him, especially as a way to reach a broader audience. From where he sits, he believes that opposing action to cut carbon pollution is best for

Australia's future. Frankly, this is incomprehensible to me, but he actually does. 'We can agree that we love our country and that we want to protect it. We want to leave clean air, clean water and clean soil for the next generation and we don't want to take huge risks,' I say. 'I think Nick and I would agree on that.' We just differ on the best approach to meet these goals.

How do we make progress if differing factions never talk to each other? Nick is a bit of an extreme example—it was always a long shot for him to change his mind—but there are people out there genuinely just looking for more information. They want to cut through all the rhetoric and just understand what's happening to their world. And if environmentalists like me come across as angry and unwilling to engage, it's our fault if the public turns away.

The first step is to acknowledge the *loss* that accepting climate science brings with it for those who hold a strong free-market ideology and oppose corporate regulation. Despite being few in number, this group is politically and economically powerful. This is why they've been so effective at delaying action on climate change. As I mentioned earlier, the journalist Naomi Klein recently attended a gathering of the Heartland Institute in 2011, and found a lot of these fears on display. In her piece for *The Nation*, Klein wrote that for the audience assembled at Heartland's conference, 'climate change is a Trojan horse designed to abolish capitalism and replace it with some kind of eco-socialism'. One of the conference speakers, Larry Bell, had previously summed up this view succinctly. He'd written that climate change 'has little to do with the state of the environment and much to do with shackling capitalism and transforming the American way of life in the interests of global wealth redistribution'.

In order to make progress on solving climate change we have to *acknowledge* that for people like Nick, the implications of accepting climate science are intellectually cataclysmic. But then we have to address their concerns head on in a sensible, rational way. Does solving climate change mean changing the way our economy works? Yes. Does it mean more regulation for polluting industries? Yes. Does it mean encouraging the development of a renewable energy sector? Yes. Does it lead to the sceptics' worst nightmare scenario of an eco-socialist society ruled by a big brother–style world government? No, that is unlikely. As we heard from Zac Goldsmith, it's possible for countries to make major progress on cutting carbon pollution without collapsing into economic or political chaos.

In the words of Naomi Klein, climate change presents some 'profoundly challenging revelations for all of us raised on Enlightenment ideals of progress, unaccustomed to having our ambitions confined by natural boundaries. This is true for the statist left as well as the neoliberal right.' Understanding what values and practices we have to let go of in order to make progress must be our starting point for working through these revelations. It won't be easy, but we need to figure out what a twenty-first century economy looks like for Australia—one that *doesn't* destroy our only home, planet earth.

'Many of our culture's most cherished ideas are no longer viable,' writes Klein. Endless growth, unchecked consumerism, a dig-it-up-and-export-it economy, measuring progress by GDP alone—climate change means that all these once-sacred canons are up for review. And there's that ticking clock towards climate tipping points. We have limited time to work through our differences before tackling the devastating impacts of global climate change. So we better start talking to one another honestly.

In that vein, maybe Mike has worked some kind of magic. By the end of our discussion the three of us are finishing each other's sentences.

ME: *Let's figure out something we can do on energy efficiency.*

NICK: *I'd love to pursue that with you.*

ME: *We should do a TV ad or something.*

NICK: *I'd love to pursue that with you because I do share that view.*

ME: *There's just too much waste of energy, everyone can agree on that.*

NICK: *I've been going to the US for 40 years, but just to be there the last two weeks is a reminder of how incredibly wasteful that society is, unnecessarily.*

ME: *Because, yeah, ultimately energy is a resource that's going to get more expensive and—*

MIKE: *—and there's, you know, great competitive economic advantage in reducing waste if you can do it.*

It looks like Nick and I have finally found some common ground we can both get excited about.

Chapter Twenty-one

COMING HOME

I'M SEATED NEXT TO A RUSSIAN BUSINESSMAN ON THE FLIGHT FROM London to Sydney. He lovingly cradles a bottle of vodka he purchased in duty free. Between regaling me with anecdotes about his work-mates, he drinks the vodka at an alarming rate. After an hour or so he falls into a heavy sleep, which gives me quiet time to write two speeches. One I'll deliver in a few days to a thousand or so volunteers at an Australian Youth Climate Coalition conference in Brisbane. The other is for a much smaller audience, but also important: my wedding vows.

There's nothing like the thick silence of a darkened plane cabin in the middle of the night to help you reflect on what's important. Thoughts and memories swirl around as I write. Simon's and my history is inextricably bound to the youth climate movement in Australia. We organised the founding summit for the AYCC together—he did the budget and logistics and anything else that required maths, and I did the relationship building to get the right people to agree to be in the room at the same time. So when Simon and I met in person for the first time, in the lobby of the dodgy Melbourne youth hostel we'd booked everyone into, I'd already spoken to him on the phone for hours.

But it hadn't prepared me for meeting this incredibly smart, confident 20-year-old man who had gone through more hard times in his life than most people deem possible. I was almost two years older, but he was taller and somehow wiser. My politics were more radical—he was an economist, after all. We facilitated most of the sessions at the founding summit together and stayed up late each night drafting the policy declaration that bound AYCC's initial member groups together. We supported each other through the rough bits and did joint media interviews at the end. When the three-day summit was over and we were still there picking up rubbish and washing the coffee mugs, I knew we'd be friends for life. We balanced each other out, made something bigger than the sum of our parts. But back then I had no idea that almost five years later to the day, he would become my husband.

As the youth climate coalition gained momentum after the summit, Simon and I spent more time together. I was AYCC's national director and we were both on the steering committee. AYCC's membership swelled from a handful of volunteers to thousands of people joining through our website. At first, I had a Google alert set up for every time someone filled out the membership form. But after a month or so I had to turn it off. My inbox was overflowing with the names of thousands of people who wanted to be part of this new movement.

Now, AYCC has over 70,000 members from every corner of the country and unites all thirty major Australian youth organisations. Eighty local grassroots groups deliver strategic, media-savvy campaigns and projects with energy and efficiency. Our handful of staff is aged under thirty, and most are in their early twenties. I am so proud of what we've built. The generation most affected by climate change is finally demanding action—and succeeding in changing hearts, minds and votes. Even over the course of a few years, I've seen our movement grow more powerful. We've built a political and economic constituency that's turned into a force to be reckoned with.

This is what awaits me, as our plane starts its descent into Sydney. I'm coming home to a man and a movement, and I love both deeply. We land, and Simon's there waiting with a huge bunch of roses and an elated smile. I sink into his hug, relieved to be home.

•

I have a few days in Sydney to shake off the jetlag before our final leg of the trip, to Heron Island. I have to work, but at the back of my mind the seed of a thought is beginning to grow. Considering what I've seen over the last four weeks, I suspect I need to quit my day job and find a way to return to full-time climate change campaigning of some sort. About half my clients at the communications agency where I work are environment or climate change groups. And chairing AYCC's board takes up most Saturdays. But half my time might not be enough when there are still so many Australians out there who don't yet understand or support a price on carbon pollution. I don't blame them—I blame the media coverage, which has, of late, been overwhelmingly negative. But the more people getting the facts out there, the better. I decide to think about it over the next few days.

And then, sooner than seems possible, I'm back at Sydney airport, this time flying to Brisbane. It's the day before AYCC's Power Shift conference and the first time I've been to Brisbane since last summer's devastating floods. When I land, the sun is shining with such fierceness that I almost imagine it's trying to prove its worth as an energy source. I'm carrying important cargo: kilograms of chocolate for the volunteers. I take it straight from the airport to AYCC's temporary office space in South Bank, making my way up a flight of stairs through swarms of young activists.

Our entire national team of staff (and a large group of volunteers from Sydney and Melbourne) all moved to Brisbane a few weeks ago for the final recruitment stretch for the three-day conference. They're living in a big house together and working out of donated office space at the Queensland Public Sector Union. I practise my speech in the kitchen in front of a group of 17-year-olds taking a break from stuffing conference bags. They clap enthusiastically, even though I'm coming down with a cold and my voice is husky.

The next morning I make my way to the high school that's hosting the conference, joining a stream of young people high on hope and shared purpose. Nick Minchin is already there, wandering around talking to some of our volunteers. I'm feeling sick and sleep-deprived, but excited to be able to share part of my world with him. The MC acknowledges Nick in the introductions, and the audience claps. They're full of love towards the whole world, even climate sceptics. Either that, or they're too young to know who Nick Minchin is.

I deliver my speech to the opening plenary while Max films and Nick listens in the front row. Here's an extract:

We've already changed the acidity of the oceans and the composition of the atmosphere. The world we see at the end of our lifetimes is going to be a very different place from the world we were born into.

We are the last generation with the ability, in terms of the time-frame, to stop the worst impacts from happening. So we must fight.

Past generations fought for us.

Our great grandparents fought to give women the right to vote, and Australia was the second country in the world to pass this into law.

Our grandparents' generation fought World War Two, so that we would be able to grow up in a world free from fascism.

Our parents' generation did everything they could to prevent nuclear war and stop two great superpowers from destroying our futures.

Past generations did all this for us.

What legacy will our generation leave? Now it's our time.

The odds are against us—and if you understand how bad the science is you might be tempted to bet that we would lose. But as Bill McKibben says, that's not a bet that you're allowed to make.

The only thing that a morally awake person can do when the worst thing that can happen is happening on your watch is to do everything you can to try to change those odds.

There is no guarantee that we will win, but we need to guarantee to each other that we will fight as if it's our future on the line.

So let's show the world that because we love this country, we can't settle for the world as it is. The future will not be written for us. It will be written by us.

•

After I've delivered my speech, Nick, Max and I rush back to Brisbane airport. We have a flight to catch to Gladstone, where we'll take a helicopter to Heron Island. In the taxi, Nick tells me my speech was well-delivered, although he obviously disagreed with the content. I'm glad he got to see it regardless, and that he got to see a thousand-plus of AYCC's volunteers who are working to create a renewable energy–powered, safe-climate future.

On the plane to Gladstone, I'm nauseated and my stomach starts convulsing. I'm getting a fever and begin to shiver under my cardigan. By the time we land, to overcast skies and a warm wind, I've deteriorated even further. In no state to make up my mind, I let Max and Nick come to a decision for me—I'll go to a doctor in Gladstone and all of us will stay here overnight. We'll fly to Heron Island tomorrow morning.

It turns out I have tonsillitis, and probably food poisoning too. The doctor prescribes the one thing that isn't possible for the next few days—rest. But I take his advice while I have the chance and collapse on the bed in my hotel room at five in the afternoon. Three hours later, I wake up to thunder echoing like cymbals and rain streaming in the open sliding doors.

I get up to pull the doors closed, but before I can I'm drawn out onto the balcony by a strange orange glow. Above me, lightning tears the sky apart. But in front, where the soil meets the sea, a gargantuan industrial estate is lit up like Times Square. The aluminium smelter here is the largest in Australia and the fourth largest in the world. Close by is the coal terminal, exporting Queensland coal, and with it climate change, to Asia and parts of Europe. The woman at hotel reception later tells me this coal port is set to become the biggest in the world, overtaking the one in my hometown of Newcastle.

I close the doors and crawl back under the sheet to let the fever run its way through my shaking body. *What have we done to this planet?* I keep thinking. *What have we done?*

Chapter Twenty-two

GOING, GOING, GONE . . .

THE NEXT MORNING I'M FEELING BETTER, AND I WAKE UP EARLY enough to witness a searing pink sunrise melt into a hazy sky. Nick, Max and I get into a helicopter, and we fly over crystal clear water studded with tiny coral cays and schools of turtles. From above, the ocean looks like a blue desert stretching far into the horizon. But too quickly, the flight is over and the pilot deposits us on Heron Island, 72 kilometres northeast of Gladstone. Pete, Kate and Leo are waiting at the helipad with cameras and sound gear in hand. Our half-day delay means we have a lot of filming to do.

As we walk along the sandy tree-lined track to the Heron Island research station, the numerous birds and animals we pass barely glance at us. They're not scared of humans; we're in their territory for once.

Our first order of business is a snorkelling trip on the Great Barrier Reef. Max wants to get underwater shots of Nick and me to set the scene for this part of the documentary. The blue water is deceptively calm-looking from the boat but once we plunge beneath the surface we see it teems with life. I see fish, sharks and even an elderly, majestic-looking turtle as I snorkel and dive between brightly coloured coral canyons. It's a whole other world down there. I feel like an alien intruder, wetsuit clad and clumsy in my flippers. The marine biologist we're here to meet, Professor Ove Hoegh-Guldberg, calls the reef an underwater rainforest because of the complexity and density of life down here.

After an hour or so, we reluctantly return to shore to peel off our wetsuits and change into dry clothes for filming. We regroup beside a collection of tanks, hoses and tubes that together make up the most groundbreaking coral reef experiment in the world. One of the scientists responsible for its creation is here to explain its significance. Ove Hoegh-Guldberg is the director of the Global Change Institute at the University of Queensland and a professor of marine studies. He's also the lead coordinating author of the oceans chapter in the upcoming IPCC assessment report. This prestigious role underlines Ove's eminence in global climate science—he's one of the most respected marine scientists in the world.

Pipes pump seawater from the reef into a set of tanks containing coral and marine life. Before it enters each tank, the water is heated to different temperatures to replicate various levels of climate change within the IPCC's range of predictions. 'What we've got here is an experiment that's looking at future conditions and how they will affect reefs,' explains Ove. As the experiment runs its course, it will yield information about how reef organisms and processes will respond to climate change.

'What's really important about putting these organisms into these treatments [tanks] is that we're also looking at these reef processes, which largely determine whether reefs are going to be around in time [or not],' says Ove.

The experiment measures the impact of higher temperatures on coral reefs and also the impact of ocean acidification, as Ove goes on to explain.

'One of the other properties of CO_2—in addition to warming the atmosphere and the earth and trapping heat—is that it also likes to go into sea water,' says Ove. 'CO_2 reacts with water to create a very dilute acid. And the net effect is that acidified condition then decreases the concentration of carbonate ions—and that's what makes the calcium carbonate.' Calcium carbonate is the crucial substrate that corals and many other marine organisms use to make their skeletons. 'Already since the pre-industrial period we've had a drop of about twenty-five per cent [of ocean carbonate],' he says.

'So as we go forward into the future we get to situations where we have vastly lower amounts of carbonate and we approach conditions that haven't been seen for millions of years on this planet,' says Ove. 'Of course that has many biologists concerned.'

'So it's essentially changing the pH of the ocean?' I clarify.

'That's right, yes,' says Ove. 'There's actually about a thirty per cent increase in the protons or the hydrogen ions that cause the acidification . . . it's a really big chemical change.'

Ove continues his explanation. 'Another thing that we've been noticing is that the fundamental metabolism of organisms is starting to alter. So things like photosynthesis and respiration, which are very important to ecosystems like this, are changing quite substantially.'

Talking about ocean carbon levels leads Nick and me to reminisce about our trip to measure atmospheric CO_2 levels at Mauna Loa in Hawaii. When we were there CO_2 levels were 389.2 parts per million (ppm). As of December 2011, this number is up to 391.8 ppm. NASA's James Hansen says the 'safe' upper limit of CO_2 in the atmosphere is 350 ppm, a limit we exceeded around two decades ago.

'We should all go back to when it hits 400,' says Nick.

'What, to celebrate?' I ask sarcastically.

'Well then you could have a retrospective,' says Ove. 'Who was right?'

'Remember that crazy coral reef biologist? He was wrong,' Ove jokes, referring to himself. 'Nothing wrong with the coral. That would be the best case,' he says wistfully.

Sadly, things are far from best case when it comes to coral reefs. 'One of the effects of warming of the oceans has been coral bleaching,' says Ove. 'Which is essentially corals getting sick when they get too warm, and after that they tend to die in greater numbers. This has been affecting reefs since 1979. It's not known before then.'

When you combine ocean acidification with coral bleaching, things aren't looking good for the reef. 'We might lose reefs for hundreds if not thousands of years,' says Ove. Reefs are delicate and conditions like ocean chemistry take a long time to recover, he tells us.

'Mostly when we talk about climate change we talk about the air,' I say. 'But it seems that oceans are just as important when it comes to the impacts of climate change?'

'Well, I'd argue they're even more important because over ninety per cent of the heat, the extra heat, has gone into the ocean,' Ove says. 'And so there's been this dramatic heating of the upper 700 metres of sea water around the planet. Those impacts are then translated in warmer sea temperatures. The polar regions of course are undergoing very dramatic changes—but even in the tropics, here in this particular ocean, we've seen about a 0.8 degree increase in sea temperature over the last hundred years.'

So what are the results of the work that scientists like Ove have put into studying coral reefs? 'It looks like corals can't survive more than a 2 degree hike in sea temperature before they get into circumstances where they can't replace themselves faster than bleaching will nail them,' reports Ove bleakly.

'But I think these changes that are going on in the ocean,' continues Ove, 'are telling us that maybe the life support systems for a lot of what we do . . . are starting to be threatened by changes which are extremely rapid relative to the way it's happened in the past.'

That's one of the things that Ove and his fellow biologists worry about as they meticulously record rapid and unusual changes in ecosystems. It's not just the level of CO_2 or the number on the thermometer that worries them, but the *pace* of change. Given that CO_2 lingers in the atmosphere for over a hundred years, the delay between carbon pollution being emitted and the impacts becoming visible is especially problematic.

This is not limited to ocean ecosystems. Back on the mainland, for example, in Queensland's wet tropics rainforests like the Daintree, unexpected changes are taking place. Many of the species living in these forests can only survive within a narrow temperature range.

For instance, the white lemuroid possum, which lives in the higher altitude rainforests of far north Queensland, doesn't sweat, doesn't pant and doesn't lick its fur. So when it gets hot, it has no way to cool down. For these possums, 'adaptation' to climate change isn't possible. They die after four or five hours of their body temperature heating beyond their temperature threshold, which is around 28 to 30 degrees Celsius. In the summer of 2005, the community of white lemuroids in and around the Daintree rainforest experienced temperatures hotter than 28 degrees Celsius for twenty-five consecutive days. Since then, barely a handful have been spotted. 'It occurs to me,' writes author Jo Chandler after visiting the forests, 'that if these possums were people, they would strafe us with bullets, or drop from the trees and tear us to pieces for what we are stealing from them—their country, their sustenance, their children's future.'

As the planet warms and the rainforests dry out, biologists have watched animals like the white lemuroid possum move higher up the mountains to cooler, wetter patches of forest. But they've almost reached the top and soon there will be nowhere higher to go. A peer-reviewed study in December 2011 showed the same thing happening to tropical birds. 'Species may be damned if they move to higher elevations to keep cool and then simply run out of habitat,' said Professor Stuart Pimm from Duke University, one of the study's co-authors.

One day Queensland's ancient wet rainforests may have vanished forever, and with it a long list of animals that will never again walk the face of the earth. Not just possums, but frogs and lizards, birds and insects, plants and fish and countless other species. Some will disappear before humans have even had the chance to catalogue their existence or mourn their loss. A report published in the journal *Conservation Letters* predicts that as few as one in five of the plants and animals that currently inhabit tropical rainforests will still be there in ninety years time.

'Now some people have talked about evolution, but evolution takes time,' says Ove. 'It's a very slow process, especially with such a long-lived organism such as a coral . . . we've got to remember that the ice age transition was probably a hundred times slower than the rate of change that we're putting ecosystems through today.'

'A 0.8 [degree temperature rise] over a hundred years is too quick for coral, is that what you're telling us?' asks Nick.

'That's right,' says Ove.

'What does it mean for food security in the region?' I ask.

'There are some parts of the ocean that may actually have more robust fisheries but they're very much fewer than the number of fisheries that look like they're going to have big problems,' Ove replies. 'In the Pacific there have been studies done recently that show that climate change and overfishing really are going to threaten the protein that those nations need.'

Tucked in my bag is a report from the UN Food and Agriculture Organization summarising some of the research about climate change and food security in the Pacific. 'Given that coastal fisheries provide a significant source of food and economic security for coastal populations . . . climate change poses a serious threat to the livelihood of Pacific people,' it states.

So when Ove talks about threats to coral, he's not *just* talking about coral. As with most things when it comes to climate change, the more you look into it, the more you find that human lives are at stake too.

Humans are a very young species from the perspective of geological time. Modern humans have only been on this planet for 200,000 years, and have been burning fossil fuels on an industrial scale for just 200-odd years. Yet in the view of some scientists, those 200 years—a mere blip in the perspective of earth's history—have ruined our chances of hanging around long enough to evolve further.

'We're going to become extinct,' stated Frank Fenner, emeritus professor in microbiology at the Australian National University, in a remarkably blunt interview with *The Australian*. He is a highly respected scientist, widely lauded for working on the vaccine that eradicated smallpox. 'Whatever we do now is too late . . . We'll undergo the same fate as the people on Easter Island,' he said. 'Climate change is just at the very beginning. But we're seeing remarkable changes in the weather already . . . It's an irreversible situation. I think it's too late.'

I don't agree with Fenner that it's too late to prevent human extinction. But I am only too aware that many other species are already in deep water—pardon the pun—because of climate change.

On our dive this morning, I came face to face with a beautiful old sea turtle with a leathery neck and a moss-covered shell—a lone representative of a gentle species that evolved hundreds of *millions* of years ago. Heron Island is a turtle breeding ground. Up to 2000 turtles return each year to the spot in the sand where they first kicked their way out of their shells as little hatchlings.

Out of all the ways that climate change is threatening turtles—and there are many—one of the biggest factors is how warming sands will affect future generations of turtle embryos.

When baby turtles grow in their sand-insulated eggs, it's the temperature of the sand that determines their sex. Normal sand temperature will yield a roughly 50/50 mix of male and female turtles. But over the past few breeding seasons at Heron, the turtle hatchlings have mostly been female. And by 2070, climate models anticipate sand temperatures that would bring about a near complete feminisation of hatchlings. Female turtles will search in vain for male mates that no longer exist.

As if this wasn't bad enough, there's another impact of warming sands on turtles. The warmer the sand temperature, the weaker the hatchling turtles. When the sand gets too hot, the turtle babies basically overcook, leading to higher rates of deformities. And if the sand gets above 33 degrees, the baby turtles die before they even have the chance to hatch. Researchers at Townsville's James Cook

University predict that by 2070, sands on many islands of the Great Barrier Reef will be so scorching that eggs could not survive.

Even small rises in sand temperature have the potential to weaken the turtles as they grow inside their eggs. And when the little hatchlings push their way out of their shells and scramble across to the reef with all its predators, only the strongest make it. So if the sand is too hot, the turtles have less chance of surviving the brutal first swim through the reef. They'll still put their heads down and paddle as hard as they can. But they're likely to be too slow to avoid being eaten by the sharks waiting in the shallows.

My heart already feels battered and bruised from reading report after report about the impacts of climate change. Yet there's enough left for it to break just a little bit more when I learn about the turtles and the lemuroid possums. Our actions in the past few hundred years are killing animal species that were around for millions of years before us.

•

That night, we leave the research station to eat at the resort restaurant on the other side of the island and tune into the rugby on their TV. After dinner I wander back along the sand under thousands of twinkling stars. I wonder if people will always play sport, no matter how bad climate change gets. Maybe we'll be too busy just trying to survive, looking for higher ground like the possums and telling our children stories about the world as it once was, abundant with miracles like coral reefs and turtles.

Chapter Twenty-three

LINE IN THE SAND

THE NEXT MORNING BRINGS OUR SECOND AND FINAL DAY ON HERON Island, and our last day of filming. Max asks Nick and me to walk along a long stretch of beach and reflect on what we've learned on our journey together. He and Pete follow us, filming our discussion.

'Well, it's been a long four weeks,' I say. I think back to our first meeting on my Uncle Geoff's farm, where we didn't know each other at all.

'It has indeed,' replies Nick.

We discuss the experts we met on each side. We reflect on Nick's line-up of blogger Joanne Codling and the sceptic activist David Evans, the climatologist Richard Lindzen, the blogger and politico Marc Morano, the writer and media personality Bjørn Lomborg, the journalist Christopher Booker and the nuclear physicist Jasper Kirkby.

We recall my line-up of oceanographer Matthew England, science historian Naomi Oreskes, the US Navy's Rear Admiral David Titley, Conservative British MP Zac Goldsmith, the former head of Amnesty International Irene Khan, and the marine scientist Ove Hoegh-Guldberg, who's currently making sandwiches for our lunch back at the research station.

And then there were the 'neutral' spokespeople selected by the production team. These were the Mauna Loa scientist John Barnes, Berkeley Professor Richard Muller, Yale psychologist Anthony Leiserowitz, and British scientist Mike Hulme. Nick and I talk about Hulme and the way he's decided to search for common ground between sceptics and the mainstream science by posing the issue of climate change simply as a question of risk.

'What he kind of helped convey, I think, is that there are still some uncertainties,' I say. 'But there are some things that we *do* know, that we know pretty well. That doesn't mean we know everything.'

'And you're at a point where you say we know enough,' says Nick.

'We definitely know enough,' I agree.

'That's the grey line, I guess,' says Nick. 'Where do you decide that the uncertainties are *not* such as to . . . adopt a cautious approach to action?'

I still can't believe that Nick can't see that line was passed long ago. It's the line in the sand that separates the certainties from the uncertainties. It doesn't say there are no parts of the climate system left to learn more about—but it underscores the fact that we know enough to act. I wish Nick would apply the 'cautious approach' he so often talks about to the question of protecting the planet. Instead, he seems more focused on protecting the economy from measures designed to save the planet.

'I reckon I'm more conservative than you think,' I tell Nick. 'What I'm arguing for is pretty conservative . . . to keep this beautiful world the way it is in terms of the climate. The *radical* thing is actually to mess with this soil and this water and this atmosphere in a way that has consequences that are damaging.'

'I appreciate that's your perspective,' replies Nick.

But he isn't convinced. 'We can never stop the climate changing,' he says. 'It will always change and the environment will always change . . . From when the bloke chopped down the first tree to build a fire and heat his stove.'

'But we've gone so far in terms of the ability humans have [to impact] on the world since that time,' I say. 'Humanity's technological ability has increased so much. We didn't *know* when we started the Industrial Revolution that it would change the climate. It wasn't like some evil . . . plan to wipe out all of the species on earth. But we do know enough now to say that we *are* causing that impact.'

'Well, I mean that remains what I think is a debate, you know,' says Nick.

'But you've accepted—and I think you probably always came into this acknowledging—that humans have some impact on the climate,' I say. 'You're just saying you're not sure how much.'

'Yeah, and whether that is dangerous,' says Nick. 'You know, it might be a bit of warming is a good thing. I've always believed that the worst thing we could have is an ice age.'

'Nick, we're not going into an ice age,' I say with exasperation. He's really bringing this up *now*, in our last conversation?

'We could easily go into another ice age,' he says. 'It's almost inevitable that we will . . . Not tomorrow but I'm talking 1000 years [we] could easily.'

'Can I wind things back to the present century?' interrupts Max. He asks us to reflect on what common ground we've ended up with after four weeks of looking at the science.

'It seems like we do actually have some common ground,' I say. 'We both accept that greenhouse gases trap heat, that temperatures have been going up over a long period of time, that carbon has been going up, and that there are some negative impacts to that. What we *disagree* on is the level of risk and the kind of response [required].'

'I think you've summed it up,' says Nick.

'I reckon that's a starting point,' I say.

I'm actually quite impressed with Nick's transformation. He started out calling climate change 'the new religion of the extreme left' and a pretext to 'de-industrialise the Western world'. Over the course of our journey, Nick has affirmed that climate change is happening and that temperatures are rising. He even said to Zac Goldsmith, 'human emissions of CO_2 probably made some contribution to that'. Yesterday, talking to Ove Hoegh-Guldberg, Nick acknowledged that the majority of the scientific community accepts that CO_2 causes climate change, even though he refutes that the level of scientific agreement is as high as ninety-seven per cent ('I don't mind people talking about *majority*,' he'd said, 'but that ninety-seven per cent is the greatest fraud ever perpetrated').

Nick still disagrees on how much natural background climate variability there is. And he still has these bizarre ideas that climate change might be *good* for the world, when the overwhelming evidence says otherwise.

But still I'm impressed that he has moved a long way ahead of the sceptics that deny temperatures have even risen, those who say that the temperature record is 'corrupt' and refuse to accept that carbon pollution could be changing the climate at all. It's even arguable that he now has more in common with me than he has with Joanne Codling and David Evans. They are still stuck arguing about whether the world is warming or not, based on some photos of temperature measurement stations next to air conditioners.

With this much common ground, how is it that Nick ended up so strongly opposing action on climate change? How did he come to see it as a left–right issue? Why did he not view it like other scientific issues that have been less divisive? Naomi Oreskes discovered that the history of climate denialism, had its origins in conservative politics—but I'm determined not to let this past determine our future.

'Is there anything that I can do to help bring people like you along on a journey of starting to reduce our emissions?' I ask Nick. 'Because obviously I've been doing something wrong in the past. Obviously the environment community hasn't been able to reach out to a hundred per cent of people. So what can I do better?'

'What I think the problem has been is . . . fear mongering,' says Nick. 'Painting such an extreme picture that ordinary people just turn off. They think, well, that's just ridiculous. They're not going to be convinced by the sort of doomsday scenarios.'

'So when people hear the science that's quite scary, it's the worst place to start, you mean?' I ask. 'We should talk about the potential *best case* scenarios [in the science] as opposed to the worst?'

'Well it's better to say there are a range of reasons why it would be sensible for our communities to become less reliant on fossil fuels,' says Nick. '(a) There's a potential risk from CO_2 emissions causing warming that could be a problem; (b) it's going to exhaust anyway; (c) it produces other—'

'Health impacts,' I add.

'Yeah, it has other health impacts,' he says. 'So there is a range of good reasons why we should seek to reduce our dependence on fossil fuels—but in a way that societies can manage.'

I'm finding it slightly strange that Nick is giving me communications advice on how to get sceptics to agree to cut carbon pollution. He's not bad at it either. A lot of climate change activists are thinking the same way. Groups like 100% Renewables and Beyond Zero Emissions completely agree that we should talk about the opportunities in renewable energy, not the aspects of climate change that tend to scare people.

'Look, if we had a capacity to harness the sun in a fashion that was as economically efficient and affordable as coal or gas, yeah, I'd be doing it tomorrow,' says Nick. 'See, I'm an optimist about human kind . . . I think that humanity will go through a transition on an energy path, which I think will be logical, affordable and sensible, which will end up with most communities powered by solar.'

Nick has a theory that the world economy is naturally de-carbonising anyway, and we should just let the market do it at its own pace.

'We're on a sort of a trajectory that involves reduction in emissions per se,' Nick says. 'In terms of the nature of our energy production it seems to me stepping back and taking a wider perspective, we're on, in a sense, a natural trajectory.'

'To a low-carbon economy, you mean?' I ask.

'Yeah, I think we are,' he replies. 'It's almost a question of, do you force feed that process by sort of massive government intervention based on the dangerous global warming scenario, or not?'

'My concession . . . is that I think we're on a trajectory which is one that you would want,' Nick adds. 'But it's not going to be at the pace that you believe to be necessary and I don't think there's any way in which we can overcome that difference. You will continue to believe we just have to do it absolutely as quickly as possible.'

'But is that really a concession or any change in your position?' I ask. It doesn't seem like one to me. 'Because you're essentially saying that we shouldn't really act as if climate change was a problem at all, we should just let the market continue, do what it wants, change the energy system really slowly—if that happens naturally.'

This 'do nothing' approach has no chance whatsoever of solving climate change. Especially in light of the 2017 deadline for global emissions to peak and decline that has been put forward by the International Energy Agency (IEA) as the only way to avoid dangerous climate change.

'I'm asking can we find a way forward to acknowledge that there is a mainstream scientific consensus that says climate change is real and is urgent and we should do something about it,' I say. 'Not just let markets continue [with business as usual] hoping that maybe they will get to a low-carbon economy in time to solve it.'

Nick talks about how transitioning from fossil fuels too quickly could end up damaging the economy. 'The economic consequences of trying to do that are I think quite dangerous,' he says. 'Because if you undermine the very basis of a community's wealth and incomes and economic standard of living you just end up with, you know, quite dangerous chaos.' Nick thinks that you can only tackle climate change with economic growth. 'The only societies that can afford, and will focus on, environmental issues are wealthy, successful socie- ties,' he says. 'That's been proven down through the ages. So you've got to do this [reduce emissions] on the basis of not fundamentally endangering the very foundations of our wealth.'

I don't accept Nick's argument. The oldest conservationists in the world are indigenous peoples, farmers and people who rely on the land. The richer a society gets, the more removed it gets from the natural world, and the more it's willing to trash the earth—because it forgets how deeply it's connected to it.

'Yet we have a movement of people around the world even in . . . countries in Africa, and India and China, saying that this is a global issue, we've got to act, we can't wait until we're super wealthy or it will be too late,' I say.

But Nick does bring up a good point. There are both economic opportunities *and* risks involved in cutting carbon pollution. The transition from fossil fuels to renewable energy is primarily an economic one. It requires both regulation and market solutions such as price signals to make dirty energy more expensive and clean energy cheaper. Given this, it's important to have people who understand markets help with the transition.

'That's the thing, we actually need people like you,' I say to Nick. 'Your background, your skills, your knowledge about public policy. We need people like you who know a lot about public policy to be part of the solution. Because to solve climate change—it's a big, it's a big ask. We're going to need to bring everyone along. Including you. Including people like you.'

'That's very refreshing,' says Nick. 'I'm really pleased with the approach you're taking because I think part of the problem has been that too many people take such an extremist point of view and just are so vile about anybody who doesn't agree with them.'

'I was brought up to respect people even if they're different to me, even if I disagree with them a lot,' I say. 'I don't think my *policy* suggestions are any different from the people that you would call "extremists". Maybe the difference is just that I'm willing to engage. I really do think it's important to bring along those people—and there are a lot of them—who aren't yet sure.'

For me, it's not just that it would be *nice* to have all of society, even former climate sceptics, support action to cut carbon pollution. I truly believe it's *necessary* if we're genuinely going to solve climate change. Can we make progress without bipartisanship? Yes, but every step will be hard-fought and vulnerable to being overturned. 'Solving climate change is really big,' I tell Nick. 'It's going to involve every aspect of society and we need the skills and the knowledge of people like you—and people in every field.'

'There's got to be a genuine attempt to reach out to people who are sceptical,' says Nick. 'To say look, we understand your scepticism, we accept that you find it difficult to believe that we're going to destroy the planet or whatever but surely there are good reasons why we should focus on finding, in a sensible fashion, a different energy mix.'

We talk about Australia's renewable energy potential, and the fact that fossil fuels are finite. 'One day we're going to have to find something else—so let's start that journey now,' Nick says. It's a line that could have come straight from a climate activist's handbook.

Nick has an interesting take on who is to blame for destroying the bipartisan support for climate science and solutions in Australia. The conventional wisdom is that it was Tony Abbott, but Nick suggests it could have been Kevin Rudd.

'You know, the Howard government actually went to that '07 election with an ETS,' he says. Nick is referring to the emissions trading scheme that former Prime Minister John Howard promised to implement before he lost government. 'But Rudd, in order to tip us out and get into government, just spent every day bashing us over the head saying these bums don't know anything about climate change and they're doing nothing and this is the great moral issue of our time,' says Nick. 'Well, that was just drawing a massive, massive divide between the two sides of Australian politics. Even though there was the potential for some common ground.'

It's an interesting point. Would less partisan messaging on climate change from the Labor Party have made a difference in getting people like Nick on side?

'What do you think of my proposition that [we should] bring it back to an energy debate?' asks Nick.

I pinch myself, still not really believing that Nick Minchin is giving me messaging advice on climate change.

'Yeah, I reckon that's a really sensible suggestion,' I say. 'Because a lot of people have very different reasons for wanting to move to a low carbon economy, to move to clean energy. Some people want to do it because of climate change. Some people want to kind of re-industrialise Australia [with renewables].'

Nick and I discuss geothermal, tidal, wind, solar photovoltaics, solar thermal energies and the exciting possibilities they represent for Australia's future. I decide I'm going to introduce him to my friend Pat at the Melbourne-based think tank Beyond Zero Emissions. I suspect Nick might enjoy talking to him about the possibilities for renewable energy in Australia. If fossil fuel subsidies were removed, clean energy would have a level playing field with dirty coal—and therefore enormous potential to expand.

'Not just clean but sustainable,' advises Nick. 'If you put the emphasis back on the fact that fossil fuels are finite whereas solar, for example, is infinite. And I think geothermal is probably infinite. I think that's a hook, so to speak, to get people on board.'

Another 'hook' climate advocates use to talk about moving away from coal is its health impacts. 'I grew up in a coal community, in Newcastle, surrounded by coal mines, the world's biggest coal export port,' I tell Nick. 'It has health impacts as well—asthma, respiratory illness.'

'Oh yeah, exactly, I appreciate that,' says Nick.

'As you said, there's a whole bunch of reasons to move away from coal anyway,' I say. 'It's just a matter of doing it in a way that is fair to people.'

'Coal has been the basis of the most extraordinary advance in human civilisation,' says Nick.

He's right. We don't need to trash the past to be able to move into the future. 'Yeah, and we've done incredible things because of our energy system,' I say.

'But, you know, there is a point at which it's obviously going to outlive its usefulness,' says Nick. 'And so I'm more than open to moving beyond the coal-based economy . . . in a way that doesn't actually undermine prosperity.'

We wrap up, talking a little more about the science. I tell Nick that I hope he ends up being right, although I'm confident there's almost no chance of that. 'I mean I wish I was wrong,' I say. 'I wish climate change wasn't real more than anything. I would do *anything* for it not to be.'

'Yeah, well I failed to reassure you,' says Nick.

'I ultimately do think that it's going to be young people who change this debate,' I say to Nick. The only climate sceptics I personally know who have changed their minds have done so because of prompting from their children. Their kids—my friends—have grown up with climate science as a fact of life and aren't afraid to hassle their parents every day about it. 'I think if anyone's going to change your mind it's going to be young people that that have been in your life for a long time, as opposed to someone who you've only just met and who disagrees with you', I say. 'I know so many of my friends who have changed their parents' minds on climate change.'

'Beat them into submission,' jokes Nick.

'Well they just, every night at the dinner table they're saying . . . Mum, Dad, this is why it matters to me,' I say.

'Persistence pays,' says Nick.

'It's kids that are making their parents turn off the lights and reduce energy,' I say. 'Because we learned about it in school.'

'Again, you bring it down to things like that,' advises Nick. 'Dad, don't you want to pay less electricity every month? Well, turn off the bloody light.'

'Yeah, exactly,' I agree. 'There are so many reasons [to act on climate change].'

'So Anna, it's been gratifying to me that you've genuinely and sincerely wanted to find out if there's any common ground between someone like you and me,' says Nick.

'I've always tried to approach the work that I do on climate change with the attitude that this movement is not just for greenies, it's for everybody,' I say. 'Because it's about something pretty basic which is survival of future generations. And frankly, I don't care what kind of politics you have. If you're willing to work towards a better world and hopefully get to a point where you can acknowledge the mainstream science then I'm willing to work with you.'

'We're talking about quite a big change to our society and our economy in our generation,' I continue. 'And to do that we need to bring along people who we disagree with or who are different to us and who are older than us—but whose talents and skills we need to make that transition.'

And in that moment, I decide that I *will* leave my job. I'll write a book that might help explain some of the science to people who aren't sure (that's this book!). I'll do as many public talks as I can to help reach out to communities who haven't yet heard someone properly explain what climate change means for young Australians' futures. And I'll take Nick up on his suggestion of focusing more on the opportunities in renewable energy by volunteering with the 100% Renewables grassroots community campaign. 'This movement has to be for everybody,' I say. 'It has to reach out to everybody . . . Maybe we won't change everyone's mind—but we've got to try.'

Nick and I walk back along the sand, retracing our steps to the research station. We each do final one-on-one filmed interviews with Max, and then it's all over. Relief washes over me. Finally I can relax, after weeks of scrutiny through the lenses of Max's and Pete's ever-present cameras.

I decide to go for a swim before we catch the ferry. This will be the swim, I decide, that will wash away all the grubbiness of the program. It will wash away all the disturbing views I've encountered not just about climate science but about passive smoking and evolution and asbestos and the role of science and the rights of corporations to trample over my future unimpeded.

I change into my swimsuit and walk along the beach, enjoying the white sand on my bare feet. But as I pull my dress over my head, the wind picks up and starts blowing grains of sand against my skin. The sky blackens. Dark birds circle overhead, and their cries are haunting. The wind turns into a minor gale. I'm in paradise, but suddenly it feels ominous. I have a quick dip, then return to the warmth of my towel. I sit on the beach and stare out to sea until it's time to walk to the ferry.

I didn't get my idyllic swim on a tropical island, but perhaps this is a fitting end. Our story doesn't have a happy ending—not here, not yet. Nick and I don't walk into the sunset laughing and holding hands. He does seem to have partly changed his mind, accepting that a doubling of CO_2 does cause warming of at least 1.2 degrees Celsius—but many of his previous assumptions, like that this level of warming doesn't cause harmful impacts, haven't been altered yet. And I still haven't managed to convince him to accept the existence of positive feedback mechanisms that worsen the initial warming from CO_2.

•

The ferry ride goes by in a blur. I spent most of it dozing in an uneasy sleep to the rhythm of the rocking boat. When we land we catch the bus to the airport, and then, abruptly, I realise the project is about to end.

Nick's flight is called. He stands up. This is probably the last time I'll see him until the show goes to air. I stand up too and reach my hand out to shake his. 'Pleasure travelling with you,' I say and we both smile. Normally when saying goodbye to someone I've travelled with for four weeks, I'd give them a hug. But I sense that would be too awkward for both of us. So we shake hands. Nick doesn't say goodbye to Max, Pete, Kate and Leo. Maybe he doesn't see them. Maybe, understandably, he's too tired to care.

•

The plane from Gladstone to Brisbane is as shaky as my heart. I'm having second thoughts about the show—obviously way too late, and useless at this stage. What if I came across as having compromised in my acceptance of the science? I haven't. And nothing I've said could be understood in this way. But an edit suite that folds hundreds of hours of footage into just one is more than capable of distorting someone's position. The production team's priority is making good television, and getting viewers.

The pastel pink polish I painted on my fingernails in Boston is starting to come away from my nails. I chip at it absent-mindedly, thinking about all the things I need to do for the wedding. It's only two weeks away, now. I guess I should try to grow my nails. Isn't that what brides do? Have nice nails on their wedding day?

On a scale of ice sheets and glaciers, reefs and research facilities, turtle extinctions and collapsing agriculture, my wedding doesn't seem so important any more. It will be simple. We've run out of time to make it anything else. But that's OK. It will still be the best day of my life.

We land in Brisbane slightly delayed by bad weather. The crew gets Max, Pete, Leo, Kate and me off the plane before other passengers so we can run to make our connecting flight to Sydney. The airport is filled with an unusual number of teenagers. Some of them smile at me shyly. And then I remember that AYCC's Power Shift conference would have just finished. Many of these young people are probably AYCC organisers from out of town, now heading home. I remember when I first understood what they will have come to understand this weekend. That when people unite for positive change, they can be powerful enough to actually win. It's an exhilarating lesson to learn. It was this realisation that set the course for my life so far.

On the flight to Sydney I think about Nick. I wonder how he's feeling as he flies home to Adelaide. Has our journey shifted his views at all? I'm proud of him, in a way. Proud that he admitted that climate change is happening and that humans 'probably play some part in that'. It's the most anyone could expect. He now understands the science more than most sceptics. And maybe some seeds were planted in his heart and mind that will grow over time.

EPILOGUE

ABOUT TWO WEEKS AFTER WE FINISH FILMING, I WAKE AT 5.30 AM ON a Saturday. It's a breathtakingly beautiful morning, the kind of day when Sydney puts its best foot forward and dares every other city to compete with its stunning good looks. We're on the cusp of late spring and early summer. It's the season of peonies and David Austin roses, hydrangeas and jacarandas, and the first cherries and mangoes.

Simon and I are getting married at Sydney's South Head National Park, where the water sparkles through the trees. The harbour sits on one side, the ocean on the other. To top it off, it's a sweet and sunny 26 degrees with a gentle breeze and not a cloud in the sky.

At 3 pm, I take a deep breath and take my first step down the aisle of rose petals, surrounded by my friends and family. At the end, under the shade of a banksia tree, is the man I met almost five years ago. I've been in love with him since soon after then, and plan on staying that way for the rest of my life. I reach the end and take Simon's hands. We grin at each other. This is it!

•

Three days after our wedding, Simon and I pack the Prius and go on our first road trip as newlyweds. Not our honeymoon, not yet.

First we're headed to Canberra to witness the government pass the price on carbon pollution into law. We've been working towards this day, at times separately and at times together, for years.

We arrive and sit together in the Senate gallery. It's packed full of climate activists, including many AYCC volunteers, who've come to witness the vote. We watch Senator Penny Wong deliver her speech amid taunts and insults from the Liberal and National Party senators. Corey Bernardi, Nick Minchin's South Australian protégé, is irate. His face contorts into a sneer when government senators speak. His colleagues yell over the top of Senator Wong. I wonder how the debate on an issue as important as climate change has been reduced to something resembling a schoolyard fight.

Penny Wong does a good job of ignoring the hecklers. She delivers a strong speech, calling the legislation 'a historic moment' and 'a reform that is long overdue':

> It is important to remember why we are doing this and how this reform speaks to us about our responsibilities. Because this is a reform not just for today but for tomorrow. It is a reform for our children. It is a reform that is about listening to the next generation and ensuring that their voices are not drowned out by the din of opposition and vested interests that we have seen on display today and in the months previously. It is a reform which recognises our need to act for the future.

I think of the long road that the Australian people have walked to get to this point. Despite thirty-seven parliamentary inquiries on how to respond to climate change since 1991, our politicians haven't always taken the issue seriously. When Senator Bob Brown first brought up global warming and sea level rise in the Senate back in 1996, his Labor and Liberal colleagues laughed at him.

The journalists in the Senate press gallery are busy writing stories about the reactions from various stakeholders. The *Daily Telegraph*'s Piers Akerman pens a piece entitled 'The New Dark Age Begins Today'. According to him, the vote I'm witnessing right now will be remembered as putting 'the nation into reverse'. He furiously tells his readers that today 'will mark Day One of Year Zero for the Australian economy.' It will forever be remembered, so he says, as 'the day when Australians were whacked with an artificial impost designed to handicap their growth, reject wealth creation and mortgage the future of their children to placate ideologically driven Green cultists.' I'm guessing he counts me as one of the 'cultists' supporting the move. But there's more: 'This is the day when the momentum of history went backward,' he says. 'This is the day the Western tradition of science-backed advancement of the human condition was rejected in favour of paganism.' Paganism? Dark Age? The end of history? And Nick Minchin calls *environmentalists* the alarmists.

I have a feeling that, in twenty years time, people will look back on the arguments made by people like Piers Akerman and shake their heads in bewilderment. But for now, sitting in the Senate gallery gazing down on the angry debate below, I just feel relieved that Australia will finally join the international move towards responsible climate action. We join the thirty-one European countries that have had a carbon price since 2005. We join the twenty-three US states that have a price on carbon. We join New Zealand with its emissions trading scheme, India with its coal tax and China with its five low carbon provinces. The world started to act on climate change long ago, and now Australia is catching up.

•

A few weeks after our trip to Canberra, Simon and I finally went on our honeymoon—a road trip to Byron Bay via the stunning national parks and state forests that New South Wales is blessed with. We drove past the majestic Stockton Bight sand dunes, stayed in a solar-powered cabin deep in the rainforest outside Dorrigo, and swam on the beaches of the beautiful north coast. We saw golden yellow butterflies flit through shades of green eucalypt forest and ate fresh-picked mangoes and macadamias on the side of the road. I felt so lucky to live in a country as beautiful and diverse as Australia, and to be able to explore it with the person I'm going to spend the rest of my life with.

I worked on this book all through our honeymoon (poor Simon!) and over summer, finishing it in two and a half months. Some of my friends told me I was crazy to write a book so quickly. But we haven't got much time left before the International Energy Agency's 2017 deadline for global emissions to start declining, and I don't intend to waste a single second.

As I write this, my uncle's farm in Moree—the place Nick and I started our journey together—has been cut off by severe flooding. Crops are damaged; more than 2200 people are sleeping in evacuation centres; and another 10,000 people are cut off from roads. The premier of New South Wales, Barry O'Farrell, described the scene as more like Venice than rural New South Wales, and said surrounding areas had been turned into 'an inland sea'. If floods like this are indeed worse than they would have been without human-induced climate change (which is what climate models tell us) it's not due to carbon pollution emitted in the past year or even in the past decade. Since carbon pollution lasts in the atmosphere for over a hundred years, the climate change impacts we experience today stem partly from pollution emitted back in 1912. And in the same way, it's not us who'll directly feel all the impacts of the carbon pollution Australia emits this year—it's our children and grandchildren.

It's this time lag that motivates the urgency you may have sensed within the pages of this book. The best time to act was yesterday, but the second best is today. Australians simply cannot keep dragging their feet like they've done for decades. It would be a tragedy if we decided to ignore the entire field of mainstream science and instead let ourselves be tricked and derailed by a handful of people, most of whom are not practising climate scientists.

We have the technology needed to solve the problem—now we just have to build the political will. This starts with helping people understand the problem our society is facing in a productive and compassionate way. We are in a crisis and there are people out there whose views are causing great harm. But in a crisis we need to turn *to* each other, not *on* each other.

I hope this book plays a small part in helping do that.

•

I wonder how Nick has reflected on our time together. Is he going to stick to what he said in London, that climate change is happening and that human emissions of greenhouse gases are probably responsible for at least part of it?

The journey we went on re-ignited the urgency I felt when I first learned about climate change. I saw barely concealed panic in the eyes of leading experts. I heard about the likelihood of sea level rise changing the map of the world. I looked at report after report predicting more extreme weather events and huge impacts on human health. I was reminded that climate change affects our agriculture, our infrastructure and global security, and lives in every country.

After all this, I can't help but conclude that doubt and delay are no longer acceptable positions for those calling themselves leaders. The evidence is too strong and the risks far too great.

People like you and me can start by helping people in our own lives understand and support the science and the solutions. We're the last generation with the chance to stop our climate system passing tipping points. So at the very least, let's try.

What You Can Do

There is no doubt in the mainstream scientific community about the reality of man-made climate change. Our challenge now is to avoid the climate change impacts we can't manage, and manage the impacts that we can't avoid.

Remember, we don't need anything other than what we already have to do this. Our minds. Our bodies. Our words. And most importantly, our hearts. You don't need to become a scientist to make a difference. The science is already clear. And you don't need to become a politician. Politicians only ever soar as high as their constituents demand. This battle will be won in the living rooms of Australian homes and around the water coolers in our workplaces. Right now, most political leaders think that supporting the price on carbon pollution will hurt them electorally. So if every reader of this book persuades just five friends or colleagues to support action to reduce carbon pollution, and those five people make their support known by calling their local politician's office, this would make a difference.

If you started reading this book unsure, about the science of climate change and now feel convinced by the evidence, please share your story with others. You can do so using the website *www.madlands.com.au*.

If you started reading this book with an interest in climate change and feel compelled to do something about it, I urge you to take your next step. I don't need to tell you what this is—you probably already know. But whatever this step is, please take it today. The world can't wait.

Some ideas for those of you wondering where to start:

▶ Help the Australian Youth Climate Coalition (*www.aycc.org.au*). If you're under 30, become an online member (it's free). If you can spare a few hours a week, join one of our eighty-plus local groups to get involved in campaigns with other young people. If you're over 30, please consider donating to help grow the youth climate movement.

You'll be helping build a group that focuses on short-term policy change and long-term cultural change so we can to solve the climate crisis. It makes sense that the generation most affected should be empowered to help solve it. What an investment in your children's or grandchildren's future! Better than a trust fund.

▶ Get involved in the 100% Renewables campaign and join your local climate action group. To find the group in your area, or for tips on starting your own group, go to *www.100percent.org.au*.

▶ Help spread information about climate science and solutions to the people you know who remain to be convinced. There's reference material available at *www.madlands.com.au* that might help—but the most important thing is to speak up. My mum, a retired teacher, has no special scientific expertise. But over the last five years she's gently and patiently talked to everyone in her life, from her friends to her hairdresser, about climate science. She's probably changed more people's minds on the issue than any other individual I know. What a hero!

▶ Stay informed. This way you can easily drop a 'did you hear about . . .' into your conversations. Some good websites to keep you up-to-date include *www.climatecommission.gov.au*, *www.skepticalscience.com* and *www.climateprogress.org*. Other credible sources are the CSIRO and the Australian Academy of Science.

We stand at a crossroads in the history of the planet. We can choose to close this book, decide it's all too much, and walk away from the silent screams of our struggling planet. Or we can join those walking the path towards a safe climate future and bring as many of our friends and family with us as possible. If enough of us choose this path, we can keep the clean air, soil and water that past generations took for granted. But as my parents used to say—the house doesn't clean itself! It really is up to ordinary people like you and me to save the world.

Let's get to work!

NOTES

Chapter 1   Minchin Impossible

page 7: The Royal Society, 'Royal Society and ExxonMobil', 4 September 2006, http://royalsociety.org/policy/publications/2006/royal-society-exxonmobil/

page 7: Guy Pearse, interviewed on 'The Greenhouse Mafia', *Four Corners*, ABC TV, 13 February 2006

page 7: Clive Hamilton, 'Climate Denial Versus Climate Science', speech delivered at the University of Queensland Global Change Institute, 24 March 2010, www.clivehamilton.net.au/cms/index.php?page=speeches, p 1

page 8: Professor David Karoly, interviewed on ABC Four Corners, 'Malcolm and the Malcontents', *Four Corners*, ABC TV, 9 November 2009

Chapter 2   Trouble on the Farm

page 14: Ross Garnaut, *2008 Climate Change Review*, available at www.garnautreview.org.au

page 23: Nick Minchin, interviewed on 'Malcolm and the Malcontents', *Four Corners*, ABC TV, 9 November 2009

Chapter 3   When Worlds Collide

page 25: Paul Barry, 'Political Fixers: Nick Minchin', *The Power Index*, www.thepowerindex.com.au/political-fixers/nick-minchin

page 26: Clive Hamilton, *Requiem for a Species*, Allen & Unwin, Sydney, 2010, pp 95–133

page 26: Aaron McCright and Riley Dunlap, 'Cool Dudes: The Denial of Climate Change Among Conservative White Males in the United States', *Global Environmental Change*, 2011, www.sciencedirect.com/science/article/pii/S095937801100104X

page 26 '*Polls show the breakdown is similar in Australia . . .*': Ross Garnaut et al, 'Australians' Views On Climate Change: Executive Summary', *Garnaut Climate Change Review 2011 Update*, www.garnautreview.org.au/update-2011/commissioned-work/australians-view-of-climate-change.htm

pages 26–7: American Psychological Association Task Force on the Interface Between Psychology and Global Climate Change, *Psychology and Global*

Climate Change: Addressing a Multi-faceted Phenomenon and Set of Challenges: A Report by the American Psychological Association's Task Force on the Interface Between Psychology and Global Climate Change, p 26, www.apa.org/science/ about/publications/climate-change.aspx

page 27: Aaron McCright, quoted in Julia Piper, 'Why Conservative White Males are More Likely to be Climate Skeptics', *Scientific American*, 5 October 2011, www.scientificamerican.com/article.cfm?id=why-conservative-white-males-are-more-likely-climate-skeptics

page 28: Stephan Lewandowsky, 'Why Do People Reject Science? Here's Why . . .', *The Conversation*, 11 November 2011, http://theconversation.edu.au/ why-do-people-reject-science-heres-why-4050

page 29: Commonwealth of Australia, Senate, 21 June 2001, Senator Nick Minchin, reported in *Crikey*, http://blogs.crikey.com.au/thestump/2011/06/22/ minchin-delivers-final-senate-speech/

Chapter 4   Pilgrimage to Perth

page 37: Temperature measurements from the four scientific agencies measuring temperatures (NASA, NOAA, the Hadley Centre and the Japanese scientific agency) all show that the last decade is the hottest since temperature records began. See http://earthobservatory.nasa.gov/IOTD/view.php?id=48574

page 37: BBC News, 'At a Glance: IPCC Report', 2 February 2007, http://news. bbc.co.uk/2/hi/science/nature/6324029.stm

page 38: NASA records show the world has warmed 0.8 Celsius since 1880. NASA Earth Observatory, http://earthobservatory.nasa.gov/Features/WorldOfChange/ decadaltemp.php. Global average CO_2 concentrations have risen from 280 ppm in 1880 to 392 ppm in 2011, a forty per cent increase. See http://co2now.org

page 38: Munich Re, 'Press Release: Two Months to Cancun Climate Summit/ Large Number of Weather Extremes As Strong Indication of Climate Change', 27 September 2010, www.munichre.com/en/media_relations/press_ releases/2010/2010_09_27_press_release.aspx

page 39: James Hansen, *Storms of My Grandchildren*, Bloomsbury, London, 2011, p 9 and chapter 3

page 40: James Hansen, *Storms of My Grandchildren*, pp ix, 296

pages 40–1: Jo Chandler, *Feeling the Heat*, Melbourne University Press, Melbourne, 2011, p 4

page 42: World Meteorological Organization, *WMO Greenhouse Gas Bulletin*, no 7, 21 November 2011, www.wmo.int/pages/prog/arep/gaw/ghg/GHGbulletin.html

page 42: 'Press Release 934: Greenhouse Gas Concentrations Continue Climbing', 21 November 2011, www.wmo.int/pages/mediacentre/press_releases/ pr_934_en.html

page 42: M Molina et al., 'Reducing Abrupt Climate Change Risk Using the Montreal Protocol and Other Regulatory Actions to Complement Cuts in CO2 Emissions', *Proceedings of the National Academy of Sciences of the United States of America*, vol 106, no 49, 8 December 2009, p 1

page 42: Hansen, *Storms of My Grandchildren*, p 285

page 43: Until 2009, many scientists believed that stabilising emissions at 450 ppm was the 'safe' level. See Lauren Morello, 'Is 350 the New 450 When it Comes to Capping Carbon Emissions?', *New York Times*, 28 September 2009, www. nytimes.com/cwire/2009/09/28/28climatewire-is-350-the-new-450-when-it-comes-to-capping-c-6627.html

page 43: 'World Risks Climate Catastrophe: IEA', *Sydney Morning Herald*, 13 December 2011, www.smh.com.au/environment/climate-change/world-risks-climate-catastrophe-iea-20111213-1os25.html

page 43: Kevin Anderson and Alice Bows, 'Reframing the Climate Change Challenge in Light of Post-2000 Emission Trends', *Philosophical Transactions of the Royal Society* vol. 366 no. 1882, 13 Nov 2008, http://rsta. royalsocietypublishing.org/content/366/1882/3863.full

pages 45–6: Jo Nova [Joanne Codling], *The Skeptics Handbook* and *Global Bullies Want Your Money*, http://joannenova.com.au/global-warming/

page 46: Greenpeace USA, 'Factsheet: Heartland Institute, Heartland', www. exxonsecrets.org/html/orgfactsheet.php?id=41

page 47: Naomi Klein, 'Capitalism vs. the Climate, *The Nation*, 28 November 2011, www.thenation.com/article/164497/capitalism-vs-climate

page 47: Jo Nova [Joanne Codling] 'Inflation: The Creeping Hand of Corruption', May 2011, http://joannenova.com.au/2010/12/inflation-video-goes-viral/ (viewed 20 January 2012)

Chapter 5   Kitchen Table Science

page 50: Jo Nova [Joanne Codling], 'There is no Saving the ABC—We want 60% of our Billion Back', 22 October 2011, http://joannenova.com.au/2011/10/there-is-no-saving-the-abc-we-want-60-of-our-billion-back/ (viewed 10 December 2011)

page 54 '*A global average increase of 1.2 degrees . . .*': Gwynne Dyer, *Climate Wars*, 2008, Scribe, Melbourne, p 13

page 54 '*Even without feedback mechanisms . . .*': Professor Matthew England, UNSW Climate Change Research Centre, personal communication, 30 January 2012

page 54: Richard Lindzen explains why 'the models are right and there's something wrong with the data' when it comes to the so-called 'hotspot'. See Denis Rancourt, 'Live Radio Interview with Richard Lindzen', *Climate Guy*, 8 July 2011, http://climateguy.blogspot.com.au/2011/07/live-radio-interview-with-richard.html

page 55: JR Petit, J Jouzel et al, 'Climate and Atmospheric History of the Past 420,000 Years from the Vostok Ice Core, Antarctica', *Nature*, 399, 1999, pp 429–36. Also Dave Schumaker, 'Climate Change: How We Know What We Know', *Geology News Blog*, 30 January 2009

page 57 '*the earth could not have moved out of the last ice age . . .*': Matthew England, UNSW Climate Change Research Centre, personal communication, 19 September 2011

pages 57–8: Australian Academy of Sciences, *The Science of Climate Change: Questions and Answers*, AAS, Canberra, 2010, p 11, www.science.org.au/reports/climatechange2010.pdf

page 59: Jo Nova [Joanne Codling], *Global Bullies Want Your Money*, p 18 http://joannenova.com.au/2009/12/global-bullies-want-your-money/

page 60: Jo Nova [Joanne Codling], 'The Main "Cause" of Global Warming is Air Conditioners', *The Skeptics Handbook*, 2009, p 7, http://jonova.s3.amazonaws.com/sh1/the_skeptics_handbook_2-3_lq.pdf

page 60: Anthony Watts, 'Is the U.S. Surface Temperature Record Reliable?', The Heartland Institute, 2009, http://wattsupwiththat.files.wordpress.com/2009/05/surfacestationsreport_spring09.pdf

pages 60–1: Australian Bureau of Meteorology, *Australian High-Quality Climate Site Networks*, 5 January 2011, www.bom.gov.au/climate/change/hqsites/

page 64: Jo Nova [Joanne Codling], *Skeptics Handbook*, p 2

page 64: Intergovernmental Panel on Climate Change (IPCC), 'Summary for Policymakers', in CB Field et al. (eds), *Intergovernmental Panel on Climate Change Special Report on Managing the Risks of Extreme Events and Disasters to Advance Climate Change Adaptation*, Cambridge University Press, Cambridge, UK, 2011, p 10

page 65: Naomi Oreskes and Erik Conway, *Merchants of Doubt: How a Handful of Scientists Obscured the Truth on Issues from Tobacco Smoke to Global Warming*, Bloomsbury, New York, 2010, p 272

page 66: Sir Nicholas Stern, 'Stern Review: The Economics of Climate Change', 2007, Cambridge University Press, Cambridge, 2007, Executive summary, pp 2 and 15

page 67: Comment by 'Frank Johnstone' on Jo Nova, 'Climate Alarmists Might Just be Captive to Basic Emotions', 23 December 2011, http://joannenova.com.au/2011/12/climate-alarmists-might-just-be-captive-to-basic-emotions/#comments (viewed 10 January 2012)

page 67: Jo Chandler, *Feeling the Heat*, p 185

page 68: Brendan Nicholson and Lauren Wilson, 'Climate Anger Dangerous, says German Physicist', *The Australian*, 16 July 2011, www.theaustralian.com.au/news/nation/climate-anger-dangerous-says-german-physicist/story-e6frg6nf-1226095587105

page 69: Jo Nova [Joanne Codling], 'Death Threats are Never OK, but for those Without Morals they can be a Useful PR Tool', http://joannenova.com.au/2011/06/death-threats-are-never-ok-but-for-those-without-morals-they-can-be-a-useful-pr-tool/ (viewed 20 January 2012)

page 71 'undermine the credibility of the establishment climate scientists': 'Former Australian Greenhouse Scientist Exposes Climate Corruption', *The New Zealand Climate Science Coalition: Commonsense About Climate Change*, 15 November 2010 http://nzclimatescience.net/index.php?option=com_content&task=view&id=684&Itemid=1

Chapter 6 Measurements in the Sky

page 74: Amy Westervelt, 'Hawaii: Our Very Own Island Nation, Battling Climate Change Via Innovation', *Forbes*, 2 September 2011, www.forbes.com/sites/amywestervelt/2011/12/29/hawaii-our-very-own-island-nation-battling-climate-change-via-innovation

page 75: The Tyndall Centre for Climate Change Research, 'Who was John Tyndall?', www.tyndall.ac.uk/About/Who-was-John-Tyndall

page 75: NASA, 'Svante Arrhenius (1859–1927)', *Earth Observatory Features*, http://earthobservatory.nasa.gov/Features/Arrhenius/arrhenius_2.php; and Svante Arrhenius, 'On the Influence of Carbonic Acid in the Air upon the Temperature of the Ground', *Philosophical Magazine and Journal of Science*, series 5, vol 41, April 1896, pp 237–76

page 76: Mauna Loa Observatory, 'Mauna Loa Observatory, Big Island of Hawaii: A Baseline Observatory of the Earth System Research Laboratory – Global Monitoring Division', brochure

page 76: Scripps Institution of Oceanography, 'Charles David Keeling Biography', http://scrippsco2.ucsd.edu/sub_program_history/charles_david_keeling_biography.html

page 76: Roger Revelle and Hans E Suess, 'Carbon Dioxide Exchange Between Atmosphere and Ocean and the Question of an Increase of Atmospheric CO2 during the Past Decades', *Tellus*, vol 9, issue 1, 18–27, 1957, p 19

page 81 *'The Keeling Curve has just kept going up and up . . .'*: US Environmental Protection Agency, Climate Change – Science, www.epa.gov/climatechange/science/recentac.html

page 82: Willis Eschenbach, 'Under the Volcano, Over the Volcano', *Watts Up With That*, 4 June 2010, http://wattsupwiththat.com/2010/06/04/under-the-volcano-over-the-volcano/ (viewed 1 December 2011)

page 85: University of California San Diego, 'How Aerosols Contribute to Climate Change', *Science Daily*, 19 June 2009, www.sciencedaily.com/releases/2009/06/090619203520.htm

pages 87–8: Naomi Oreskes and Erik Conway, *Merchants of Doubt*, pp 269–70

Chapter 7 Our First Climate Scientist

page 90: Austin Bradford Hill, 'The Environment and Disease: Association or Causation?' *Proceedings of the Royal Society of Medicine*, vol 58, no 5, May 1965, pp 295–300, quoted in Naomi Oreskes and Erik Conway, *Merchants of Doubt*, p 273

pages 90–1: James Hansen, 'Coal-fired Power Stations are Death Factories. Close Them', *The Guardian*, 15 February 2009, www.guardian.co.uk/commentisfree/2009/feb/15/james-hansen-power-plants-coal

page 95: James Hansen, *Storms of My Grandchildren*, p 42

page 95: *'If our current fossil fuel–intensive emissions trajectory . . .'* Intergovernmental Panel on Climate Change (IPCC), *Climate Change 2007: Synthesis Report*,

Fourth Assessment Report, www.ipcc.ch/publications_and_data/ar4/syr/en/
figure-spm-5.html

page 97: For more information see 'Climate Change: Water Vapour Makes for a
Wet Argument', *Skeptical Science*, September 2010, www.skepticalscience.com/
Climate-change-Water-vapor-makes-for-a-wet-argument.html

page 102: Dan Gilbert, 'If Only Gay Sex Caused Global Warming', *Los Angeles
Times*, 2 July 2006, http://articles.latimes.com/2006/jul/02/opinion/
op-gilbert2

Chapter 8   Small Graph, Big Trouble

page 104: Tim Flannery, *Here on Earth*, Text Publishing, Melbourne, 2010,
pp 273–4

pages 106–7: Paul Krugman, 'The Truth, Still Inconvenient', *New York Times*, 3
April 2011, www.nytimes.com/2011/04/04/opinion/04krugman

page 107: Gordon MacDonald et al., 'The Long Term Impact of Atmospheric
Carbon Dioxide on Climate', *JASON Technical Report*, JSR-78–07, Arlington,
VA, SRI International, 1979

pages 107–8: Anthony Watts, 'Briggs on Berkeley's Forthcoming BEST Surface
Temperature Record, plus my Thoughts from my Visit There', *Watts Up With
That?*, 6 March 2011, http://wattsupwiththat.com/2011/03/06/
briggs-on-berkeleys-best-plus-my-thoughts-from-my-visit-there

page 108 *'Support also came in from Koch Industries . . .'*: Toxic 100 Air Polluters
Index 2010, *Corporate Toxics Information Project*, Political Economy Research
Institute, University of Massachusetts, March 2010, www.peri.umass.edu/
toxic_index

page 108 *'Greenpeace calls the Koch Foundation . . .'*: Greenpeace USA, *Koch
Industries: Secretly Funding the Climate Denial Machine*, 30 March 2010, www.
greenpeace.org/usa/en/media-center/reports/koch-industries-secretly-fund

page 108 *'The Koch Foundations is the largest single donor . . .'*: Berkeley Earth
Surface Temperature, 'Financial Support', http://berkeleyearth.org/donors
(viewed 1 January 2012)page 108 Muller & Associates, 'List of Major
Projects', www.mullerandassociates.com/projects.php (viewed 1 January 2012)

page 112: ME Mann, RS Bradley and MK Hughes, 'Global-scale Temperature
Patterns and Climate Forcing Over the Past Six Centuries', *Nature*, 392(6678),
1998, pp 779–87, viewed at www.astro.uu.nl/~werkhvn/study/Y3_05_06/
data/talk/14-juni/mannetal1998.pdf

page 112: SourceWatch, 'Steve McIntyre', www.sourcewatch.org/index.
php?title=Steve_McIntyre

page 113: Fred Pearce, 'Climate Change Debate Overheated After Sceptic Grasped
"Hockey Stick"', *Guardian*, 9 February 2010, www.guardian.co.uk/
environment/2010/feb/09/hockey-stick-michael-mann-steve-mcintyre

page 113: National Research Council of the National Academies of Sciences,
Surface Temperature Reconstructions for the Last 2000 Years, The National
Academies Press, Washington DC, 2006

page 114: Robert Rhode, Judith Curry, Richard Muller et al., *Berkeley Earth Temperature Averaging Process*, October 2011, http://berkeleyearth.org/available-resources/

page 115: Joe Romm, 'The Koch-funded Scientist Who Came in From The Cold: Muller Warns We're in "Dangerous Realm" of "Very Steep Warming"', *Climate Progress*, 14 November 2011, http://thinkprogress.org/romm/2011/11/14/367597/the-koch-funded-scientist-who-came-in-from-the-cold-muller-warming/?mobile=nc

page 118 *'In reality the ocen is warming too'*: From 2005 to 2010 the global oceans have continued to warm. See K Von Schuckmann and PY Le Traon, How Well Can We Derive Global Ocean Indicators from ARGO data?', *Ocean Science Discuss*, vol 8, 2011, pp 999–1024, www.ocean-sci-discuss.net/8/999/2011/osd-8-999-2011.pdf

Chapter 9   San Diego: Doubt is Our Product

page 121: CS Lewis, *The Four Loves*, quoted in Oreskes, *Merchants of Doubt*, p 41

page 122: Jo Nova [Joanne Codling], *Global Bullies Want Your Money*, 2011, p 11

page 122: R Revelle, W Broecker, H Craig, CD Keeling, J Smagorinsky, *Restoring the Quality of our Environment: Report of the Environmental Pollution Panel*, President's Science Advisory Committee, The White House, Washington, DC, 1965, p 120, available at: http://dge.stanford.edu/labs/caldeiralab/Caldeira%20downloads/PSAC,%201965,%20Restoring%20the%20Quality%20of%20Our%20Environment.pdf

page 122: National Academy of Sciences of the United States of America, *An Evaluation of the Evidence for CO2 Induced Climate Change*, Assembly of Mathematical and Physical Sciences, Climate Research Board, Study Group on Carbon Dioxide, 1979

page 123: Naomi Oreskes, 'Merchants of Doubt', *Cosmos*, issue 38, May 2011, www.cosmosmagazine.com/node/4376/full

page 123: Brown and Williamson, *Smoking and Health Proposal Memo*, Tobacco Documents Online, http://tobaccodocuments.org/landman/332506.html

page 124: Naomi Oreskes, *Cosmos*

page 125: Peter Doran and Maggie Kendall Zimmermann, 'Examining the Scientific Consensus on Climate Change', *Eos, Transactions, American Geophysical Union*, vol 90 no 3, p 22, 2009, cited in Haydn Washington and John Cook, *Climate Change Denial: Heads in the Sand*, Earthscan, London, 2011, p 47

page 125: S Fred Singer and Kent Jeffries, 'The EPA and the Science of Environmental Tobacco Smoke', Alexis de Tocqueville Institution, Arlington, VA, 1993, p2, viewed at http://legacy.library.ucsf.edu/documentStore/t/i/o/tio54b00/Stio54b00.pdf

page 125: Naomi Oreskes, *Cosmos*

page 126: Report of the Senate Community Affairs References Committee, *The Tobacco Industry and the Costs of Tobacco-Related Illness*, Senate Printing Unit, Canberra, December 1995, pp 120–1

page 126: Kate Legge, 'Nick Minchin was a Sceptic on Tobacco, *The Australian*, 1 December 2009, www.theaustralian.com.au/politics/nick-minchin-was-a-sceptic-on-tobacco/story-e6frgczf-1225805535960

page 127 *'If the entire West Antarctic ice sheet melts . . .'*: James Hansen, *Storms of My Grandchildren*, p 83

pages 127–8: Arnold Schwarzenegger, quoted in United Nations Environment Program, *A Memorable Event: World Environment Day 2005*, San Francisco, www.unep.org/Documents.Multilingual/Default.asp?DocumentID=434&ArticleID=4814&l=en

page 134: Naomi Oreskes, 'Beyond the Ivory Tower: The Scientific Consensus on Climate Change', *Science*, 3 December 2004, vol 306, no 5702, p 1686

page 137: Naomi Oreskes, *Cosmos*

pages 137–8: Nicholas Stern quoted in Alison Benjamin, 'Stern: Climate Change a "Market Failure"', *Guardian*, 29 November 2007, www.guardian.co.uk/environment/2007/nov/29/climatechange.carbonemissions

pages 140–1: Margaret Atwood, *The Year of the Flood*, Bloomsbury, London, 2009, p 239

Chapter 10   Thirteen Minutes Past Midnight

page 145: Gwynne Dyer, 'Four Harsh Truths About Climatic Change', *Japan Times*, 7 December 2008, www.japantimes.co.jp/text/eo20081207gd.html

page 146: Geoffrey Luck, 'How Green is their Energy?', *Quadrant*, September 2011, vol LV, no 9, www.quadrant.org.au/magazine/issue/2011/9/how-green-is-their-energy

page 146: 'Malcolm and the Malcontents', *Four Corners*, ABC TV, broadcast 9 November 2009, www.abc.net.au/4corners/content/2009/s2737676.htm

page 146: Vàclav Klaus, 'Climate Change: The Dangerous Faith', *Quadrant*, September 2011, vol LV, no 9, www.quadrant.org.au/magazine/issue/2011/9/climate-change-the-dangerous-faith

page 148: Oxfam Australia, 'Climate Anomalies', www.oxfam.org.au/explore/climate-change/impacts-of-climate-change/climate-anomalies

page 148: Munich Re, 'Press Release: Two Months to Cancun Climate Summit/ Large Number of Weather Extremes As Strong Indication of Climate Change', 27 September 2010, www.munichre.com/en/media_relations/press_releases/2010/2010_09_27_press_release.aspx

pages 148–51: Bill McKibben, *Eaarth: Making a Life on a Tough New Planet*, Black Inc, Melbourne, 2010, pp 2–8

pages 150–1: Ross Gelbspan, 'Beyond the Point of No Return', *The Heat is Online*, www.heatisonline.org/contentserver/objecthandlers/index.cfm?ID=7203&method=full

Chapter 11   The Nutty Professor

page 153: Ross Gelbspan, 'The Heat is On: The Warming of the World's Climate Sparks a Blaze of Denial', *Harper's Magazine*, December 1995

page 155 'In fact, scientists now know for certain . . .': Professor David Griggs, Monash University, personal communication, 17 January 2012

page 158: 'Survival Strategy #9: Amateur Radio vs the Zombie Apocalypse', The Zombie Hunter, 28 January 2011, http://the-zombie-hunter.blogspot.com. au/2011/01/survival-strategy-9-amateur-radio-vs.html

page 160: Elizabeth Kolbert, Field Notes from a Catastrophe, Bloomsbury, New York, 2006, p 109

page 160 'The difference in global average temperature . . .': James Hansen and Makiko Sato, Paleoclimate Implications for Human-Made Climate Change, NASA Goddard Institute for Spaces Studies and Columbia Earth Institute, New York, 2011, p 5

page 160 'although some scientists say the range is broader . . .': E Jansen et al., 'FAQ 6.2, Chapter 6: Palaeoclimate' in Climate Change 2007: The Physical Science Basis, Contribution of Working Group I to the Fourth Assessment Report of the Intergovernmental Panel on Climate Change , eds S Solomon, et al., Cambridge University Press, Cambridge and New York

page 166: 'Debunking Joanne Nova's Skeptics Handbook Part 3', DeSmog Blog, 23 December 2008, www.desmogblog.com/ debunking-joanne-nova-climate-skeptics-handbook-part-3-climate-models-have-it-right

page 170: Richard Lindzen, 'Illusion #7, The Anti-Galileo', Skeptical Science, www. skepticalscience.com/print.php?n=730

Chapter 12   Attack Dogs: Washington DC

page 174 'The Australian reports that the centre . . .': 'Exxon Still Aids Climate Sceptics', The Australian, 20 July 2010, www.theaustralian.com.au/news/ world/exxon-still-aids-climate-sceptics/story-e6frg6so-1225894256861

page 174 'The Swift Boat smears were allegations . . .': 'Wise Counsel', ABC News (USA) Political Unit, 25 August 2004, http://web.archive.org/ web/20050417124306/http://www.abcnews.go.com/sections/politics/ TheNote/TheNote_Aug2504.html

page 174 'The New York Times opined that the Swift Boat attacks . . .': 'Opinion: Swift Boats and the Texas Nexus', New York Times, 25 August 2004, www. nytimes.com/2004/08/25/opinion/swift-boats-and-the-texas-nexus.html

page 174 '"Alarmists are attempting to enact . . . "': Senator James Inhofe, 'The Science of Climate Change', Senate Floor Statement, 28 July 2003, http:// inhofe.senate.gov/pressreleases/climate.htm

page 175 'Scheider said he had observed . . .': Leo Hickman, 'US Climate Scientists Receive Hate Mail Barrage in Wake of UEA Scandal', Guardian, 5 July 2010, www.guardian.co.uk/environment/2010/jul/05/hate-mail-climategate

page 175 'Scheider believed it was only a matter of time . . .': Stephen Leahy, 'Violent Backlash Against Climate Scientists', Tierra America: Environment and Development, www.tierramerica.info/nota.php?lang=eng&idnews=3337

page 175 'One of the Guardian's editors, Leo Hickman, wrote . . .': Leo Hickman, 'Climate Sceptic Morano's "Courage" Award is a Vicious Irony', Guardian, 13 July 2010, www.guardian.co.uk/environment/blog/2010/jul/13/climate-sceptic-morano-award

page 175 'On another occasion he told a TV news outlet . . .': Jeff Poor, 'Left on the Cutting Room Floor: Climate Depot's Marc Morano Takes on ABC News', Dan Harris, MRC Newsbusters, 25 May 2010, http://newsbusters.org/blogs/jeff-poor/2010/05/25/left-cutting-room-floor-climate-depots-marc-morano-takes-abc-news-dan-har#ixzz0oyzpqB7t

page 176: Climate Progress, '"Kill Some Crackers": GOP Group Staffed by Marc Morano Pays Fox Affiliates to Influence Election with Anti-Obama Hate Speech', Climate Progress, 2 November 2010, http://thinkprogress.org/romm/2010/11/02/206974/marc-morano-national-republican-trust-breaking-point-video-anti-obama-hate-speech/

page 179: Joe Romm, 'Scientist: "Our Conclusions were Misinterpreted" by Inhofe, CO_2—but not the Sun—"is Significantly Correlated" with Temperature since 1850', Think Progress, 12 Dec 2008, http://thinkprogress.org/romm/2008/12/12/203456/scientist-our-conclusions-were-misinterpreted-by-inhofe-co2-but-not-the-sun-is-significantly-correlated-with-temperature-since-1850/

Chapter 13   In the Navy

page 192 'An article in the New York Times . . .': Justin Gillis, 'As Permafrost Thaws, Scientists Study the Risks', New York Times, 16 December 2011, www.nytimes.com/2011/12/17/science/earth/warming-arctic-permafrost-fuels-climate-change-worries.html?_r=1

page 192 'This permafrost is now melting for the first time . . .': Elizabeth Kolbert, Field Notes From a Catastrophe, p 15

page 192 'At the end of 20111 . . .': Christian Melsheimer, Georg Heygster and Justus Notholt, Arctic Warming: Sea-ice Minimum is Not a One-Off, Correspondence, Nature, 478, 188, 13 October 2011, cited in Verity Payne, The Carbon Brief, 13 October 2011, www.carbonbrief.org/blog/2011/10/arctic-sea-ice-minimum-in-2007-not-a-one-off

page 192 'James Hansen predicts that . . .': James Hansen, Storms of My Grandchildren, p 165

page 194 'The melting unlocks the gases . . .': Justin Gillis, 'As Permafrost Thaws, Scientists Study New Risks', New York Times and K Schaefer, T Zhang, L Bruhwiler and AP Barrett, 'Amount and Timing of Permafrost Carbon Release in Response to Climate Warming', Tellus B, vol 63, issue 2, pp 165–180

page 194 'A paper from the . . .': Joe Romm, 'Carbon Time Bomb in the Arctic: New York Times Print Edition Gets the Story Right', Climate Progress, 19 December 2011, http://thinkprogress.org/romm/2011/12/19/392242/carbon-time-bomb-in-arctic-new-york-times-print-edition-gets-the-story-right/

page 194: ibid

page 195: James Hansen, *Storms of My Grandchildren*, p 164
page 195: Gwynne Dyer, *Climate Wars*, p xii
page 196: Margaret Beckett, 'Environment and Global Security', *Daily Times*, 23 April 2007, www.dailytimes.com.pk/default.asp?page=2007\04\23\ story_23–4-2007_pg3_4
pages 196–7: Gwynne Dwyer, *Climate Wars*, pp xi–xii
pages 198–9: James Hansen, *Storms of My Grandchildren*, p 168

Chapter 14   All in the Mind

page 201: 'A Matter of Degree: What's Your Climate Profile?', Yale Project on Climate Change and the George Mason University Center for Climate Change Communication, http://uw.kqed.org/climatesurvey/index-kqed.php
page 203: ibid

Chapter 15   Bjørn Again

page 211: Bjørn Lomborg, 'The True State of the Planet', *Politiken*, Denmark, 12 January 1998, cited on Lomborg Errors website, www.lomborg-errors.dk/ quotesbyLomborg.htm
page 211: Comment by 'Governmentality' on Howard Friel, 'Bjørn Lomborg's missing questions', *Guardian*, 31 August 2010, www.guardian.co.uk/ commentisfree/cif-green/2010/aug/30/lombard-missing-questions-climate-change
page 212: Juliette Jowit, 'Bjørn Lomborg: $100bn a Year Needed to Fight Climate Change', 30 August 2010, *Guardian*, www.guardian.co.uk/environment/2010/ aug/30/bjorn-lomborg-climate-change-u-turn
page 214: Muriel Boselli, 'IEA warns of Ballooning World Fossil Fuel Subsidies', Reuters, 4 October 2011, www.reuters.com/article/2011/10/04/ us-iea-idUSTRE7931CF20111004
page 215: 'World Risks Climate Catastrophe: IEA', *Sydney Morning Herald*, 13 December 2011, www.smh.com.au/environment/climate-change/world-risks-climate-catastrophe-iea-20111213–1os25.html
pages 218–19: Sharon Begley, 'Book Review: The Lomborg Deception', *Newsweek*, 21 February 2010, www.thedailybeast.com/newsweek/2010/02/21/book-review-the-lomborg-deception.html
page 219: ML Parry, OF Canziani, JP Palutikof, PJ van der Linden and CE Hanson (eds), *Contribution of Working Group II to the Fourth Assessment Report of the Intergovernmental Panel on Climate Change*, 2007,Cambridge University Press, Cambridge and New York, viewed at www.ipcc.ch/publications_and_data/ ar4/wg2/en/spmsspm-c-6-health.html

Chapter 16   The Conserve in Conservative

page 223: Cherry Wilson, 'Indian Summer Brings Out the Crowds', *Guardian*, 2 October 2011, www.guardian.co.uk/uk/2011/oct/02/indian-summer-brings-out-crowds

NOTES

page 223: Sam Jones, 'UK Weather Returns to Normal After Record-Breaking Heatwave', *Guardian*, 3 October 2011, www.guardian.co.uk/uk/2011/oct/03/uk-weather-normal-record-heatwave?newsfeed=true

pages 231–2: Commonwealth of Australia, Senate, 21 June 2001, Senator Nick Minchin, reported in *Crikey*, http://blogs.crikey.com.au/thestump/2011/06/22/minchin-delivers-final-senate-speech/

page 236: Minerals Council of Australia, 'A New Carbon Pricing Scheme', May 2011, p 9, www.mineralscouncil.com.au/file_upload/files/media_releases/CC_Submission_May11_final.pdf

page 236: Bruce Chapman, 'Explaining the Figures: Why We Shouldn't Worry About the Loss of 23,000 Mining Jobs', *The Conversation*, 6 June 2011, http://theconversation.edu.au/explaining-the-figures-why-we-shouldnt-worry-about-the-loss-of-23–000-mining-jobs-1705

page 236: Sid Maher, 'Miners Say 23,000 Jobs At Risk', *The Australian*, 16 May 2011, www.theaustralian.com.au/national-affairs/miners-say-23000-jobs-at-risk/story-fn59niix-1226056363516

pages 236–7: Commonwealth of Australia, Senate, 8 November 2001, Senator Alan Eggleston

page 237: Chapman, *The Conversation*, 6 June 2011

page 238 *'Wind, Solar and other clean technologies . . .'*: Terry Macalister, 'Green Energy Overtakes Fossil Fuel Investments, says UN', 3 June 2009, *Guardian*, www.guardian.co.uk/environment/2009/jun/03/renewables-energy

page 238 *'In 2011, the disparity had grown . . .'*: Alex Morales, 'Renewable Power Trumps Fossils for First Time as UN Talks Stall', *Bloomberg*, 26 November 2011, www.bloomberg.com/news/2011-11-25/fossil-fuels-beaten-by-renewables-for-first-time-as-climate-talks-founder.html

page 238 *'"The trillionth-dollar milestone shows . . ."'*: Jennifer Kho, 'Global Clean Energy Investments Surpass $1 Trillion', *Forbes*, 6 December 2011, www.forbes.com/sites/jenniferkho/2011/12/06/global-clean-energy-investments-surpass-1-trillion/

page 239 *'The Scottish Deparment of Energy says . . .'*: 'Scottish Renewable Electricity on Track for 'Record Year', *Guardian*, 22 December 2011, www.guardian.co.uk/environment/2011/dec/22/scottish-renewable-electricity-record-year and Department of Energy and Climate Change (UK), 'Press Notice: Energy Statistics', 22 December 2011, www.decc.gov.uk/assets/decc/11/stats/publications/energy-trends/3918-pn11-113.pdf

page 239 *'A report published in the journal* Energy Policy . . .': Joanna Zelman, '100 Percent Renewable Energy Achievable by 2030: Study', *Huffington Post*, 25 January 2011, www.huffingtonpost.com/2011/01/25/100-percent-renewable-ene_n_813256.html

page 240: Mark Z. Jacobsen and Mark A Delucchi, 'Providing all Global Energy with Wind, Water, and Solar Power, Part 1: Technologies, Energy Resources, Quantities and Areas of Infrastructure, and Materials', *Energy Policy*, vol 39, issue 3, 30 December 2010

page 241: 'In one day, the amount of sunlight reaching . . .': European Renewable
Energy Council and Greenpeace International, 'Energy [R]evolution:
A Sustainable Australia Energy Outlook', Primavera Quint Printing,
The Netherlands, January 2007, p 60, www.energyblueprint.info/fileadmin/
media/documents/energy_revolution.pdf

page 241: Matthew Wright, 'Waking Up to the Solar Dawn', Climate Spectator, 5
December 2011, www.climatespectator.com.au/commentary/waking-
solar-dawn

Chapter 17   Meddling in the Media

page 244: 'Euromyth: Pets to be Pressure Cooked', European Commission, http://
ec.europa.eu/unitedkingdom/press/euromyths/myth59_en.htm

page 244: Philip Ball, 'The Real Global Warming Disaster by Christopher Booker',
The Observer, 15 November 2009, www.guardian.co.uk/books/2009/nov/15/
real-global-warming-christopher-booker

page 247: Department of Energy and Climate Change (UK), 2050 Pathways
Analysis, December 2011, www.decc.gov.uk/en/content/cms/
tackling/2050/2050.aspx

page 248 'In 2010 the wind sector globally . . .': World Wind Energy Association,
World Wind Energy Report, 2010, www.wwindea.org/home/images/stories/
pdfs/worldwindenergyreport2010_s.pdf

page 248 'This will generate about seventeen per cent . . .': Jim Bai and Chen Aizhu,
'China Wind Power Capacity Could Reach 1,000 GW by 2050', Reuters,
19 October 2011, www.reuters.com/article/2011/10/19/us-china-power-
wind-idUSTRE79I1ED20111019

page 248 'Bloomberg Investment suggests . . .': Bloomberg News, 'China Offshore
Wind Power Capacity to Reach 30GW, Xinhua Says', 22 June 2011, www.
bloomberg.com/news/2011-06-22/china-offshore-wind-power-capacity-to-
reach-30gw-xinhua-says.html

page 248 'According to researchers at Beyond Zero Emissions . . .': University of
Melbourne Energy Research Unit and Beyond Zero Emissions, 'Zero Carbon
Australia Stationary Energy Plan', June 2010, http://beyondzeroemissions.
org/zero-carbon-australia-2020

page 249: P Hearps and D McConnell, Renewable Energy Technology Cost Review,
Melbourne Energy Institute Technical Paper Series, May 2011, www.earthsci.
unimelb.edu.au/~rogerd/Renew_Energy_Tech_Cost_Review.pdf

page 252: Richard Wilson, '"Misinformed", "Substantially Misleading" and
"Absurd": the UK Government's Verdict on Christopher Booker's Claims', 21
November 2008, http://richardwilsonauthor.wordpress.com/2008/11/21/
misinformed-substantially-misleading-and-absurd-the-uk-governments-verdict-
on-christopher-bookers-claims/

page 252: George Monbiot, 'The Superhuman Cockups of Christopher Booker', 13
May 2011, Guardian, www.guardian.co.uk/environment/
georgemonbiot/2011/oct/13/christopher-booker

page 254: Robert Manne, 'Bad News: Murdoch's Australian and the Shaping of the Nation', *Quarterly Essay*, 43, September 2011, Black Inc, Melbourne
page 255: Wendy Bacon et al., 'A Sceptical Climate: Media Coverage of Climate Change in Australia 2011, Part 1—Climate Science', Australian Centre for Independent Journalism and Global Environmental Journalism Initiative, University of Technology Sydney, Sydney, 2011 pp 7–9
page 255: UTS Newsroom, 'Newspapers Lose their Balance on Climate Coverage', 1 December 2011, http://newsroom.uts.edu.au/news/2011/12/newspapers-lose-their-balance-on-climate-coverage

Chapter 18   Of Cosmic Rays and Compassion

page 261: Richard Gray, 'Legal Bid to Stop CERN Atom Smasher From "Destroying the World"', *Telegraph* (UK), 30 August 2008, www.telegraph.co.uk/news/worldnews/europe/2650665/Legal-bid-to-stop-CERN-atom-smasher-from-destroying-the-world.html
page 262: Henrik Svensmark, 'While the Sun Sleeps', 9 September 2009, *Watts Up With That*, http://wattsupwiththat.com/2009/09/10/svensmark-global-warming-stopped-and-a-cooling-is-beginning-enjoy-global-warming-while-it-lasts/ (viewed 10 February 2010)
page 263: ibid
page 264: Bart Vergehegen, 'Aerosol Effects and Climate, Part II: The Role of Nucleation and Cosmic Rays', *Real Climate*, 15 April 2009, www.realclimate.org/index.php/archives/2009/04/aerosol-effects-and-climate-part-ii-the-role-of-nucleation-and-cosmic-rays/
page 265: 'What's the Link Between Cosmic Rays and Climate Change?', *Skeptical Science*, www.skepticalscience.com/cosmic-rays-and-global-warming-advanced.htm
page 268: Nigel Calder, 'CERN Experiment Confirms Cosmic Rays Influence Climate Change', 24 August 2011, http://wattsupwiththat.com/2011/08/24/breaking-news-cern-experiment-confirms-cosmic-rays-influence-climate-change/
page 269: Comment by Scott Mandia on the post 'Mike Hulme Sets Lawrence Solomon and Marc Morano Straight', *Deep Climate*, 15 June 2010, http://deepclimate.org/2010/06/15/mike-hulme-sets-lawrence-solomon-and-marcmorano-straight/
page 269: Steven Schneider, 'An Overview of the Climate Change Problem', Stanford University, http://stephenschneider.stanford.edu/Climate/Overview.html

Chapter 19   That Sinking Feeling

page 272: Jo Nova [Joanne Codling], 'What About the Precautionary Principle?', *The Skeptics Handbook*, 2009, p 14
page 273 *On average a Bangladeshi emits . . .*: Simon Rogers and Lisa Evans, 'The Datablog', *Guardian*, 31 January 2011, www.guardian.co.uk/news/datablog/2011/jan/31/world-carbon-dioxide-emissions-country-data-co2

page 273 *'India would lose just 0.4 per cent . . .'*: Jo Chandler, *Feeling the Heat,* p 118

page 276: World Vision Australia, *One Earth,* www.worldvision.com.au/OurWork/ Solutions/OneEarth.aspx

page 276: Martin Barillas, 'Desmond Tutu: Climate Change is Worse than Apartheid', *Spero News,* 23 November 2011, www.speroforum.com/a/ SQZTVOSOOC14/DesmondTutu-climate-change-is-worse-than-apartheid

Chapter 20 Collapsing Cliffs

page 282: Chamber of Commerce of the United States of America, 'Petition of the Chamber of Commerce of the United States of America for EPA to Conduct its Endangerment Finding Proceeding On The Record Using Administrative Procedure Act 556 and 557', Appendix 1: 'Detailed Review of EPA's Health and Welfare Scientific Evidence', p 4; available at http://motherjones.com/files/ USCOC%20comments%20to%20EPA%20June%202009%20attachment.pdf

page 282: Bill McKibben, speech at the Commonwealth Club, 8 September 2011, http://envirobeat.com/?p=3604

page 283: Mike Swain, 'Visiting the Front Line of Climate Change in Britain', *Mirror,* 3 December 2009, www.mirror.co.uk/news/uk-news/visiting-the-front-line-of-climate-change-434772

page 284: ibid

page 285: D Anthoff et al., *Global and Regional Exposure to Large Rises in Sea-level: A Sensitivity Analysis,* Tyndall Centre for Climate Change Research, Working Paper 96, October 2006, www.tyndall.ac.uk/content/global-and-regional-exposure-large-rises-sea-level-sensitivity-analysis-work-was-prepared-st

page 285: Jo Chandler, *Feeling the Heat,* p 117

page 285: Australian Government, *Risks to Coastal Settlements, Infrastructure and Ecosystems,* OzCoasts Australian Online Coastal Information, www.ozcoasts. gov.au/climate/risks.jsp

pages 285–6: Matthew Kelly, 'Rising Threat: Higher Seas Could Swamp Coal-Loaders', *Newcastle Herald,* 20 October 2011, www.theherald.com.au/news/ local/news/general/rising-threat-higher-seas-could-swamp-coalloaders/ 2329901.aspx?storypage=2

page 286: S Solomon, D Qin, M Manning, Z Chen, M Marquis, KB Averyt, M Tignor and HL Miller (eds), 'IPCC, 2007: Summary for Policymakers', *Climate Change 2007: The Physical Science Basis. Contribution of Working Group I to the Fourth Assessment Report of the Intergovernmental Panel on Climate Change,* Cambridge University Press, Cambridge and New York, Table SPM.3, p 13

page 286: CSIRO, *Sea Level Projections: Observations vs Projections,* CSIRO National Research Flagships, www.cmar.csiro.au/sealevel/sl_proj_obs_vs_proj.html

page 286: Stefan Rahmstorf and Martin Vermeer, 'Global Sea Level Linked to Global Temperature', *Proceedings of the National Academy of Sciences of the United States of America,* vol 106, no 51, 22 December 2009

page 286 *'Antarctica and Greenland together contain . . .'*: Intergovernmental Panel on
Climate Change, Working Group 1: The Scientific Basis, *IPCC Third
Assessment Report: Climate Change 2001*, Table 11.3, www.ipcc.ch/ipccreports/
tar/wg1/412.htm#tab113
pages 293–4: Geoff Aigner, *Leadership Beyond Good Intentions*, Allen & Unwin,
Sydney, 2011, pp 87–8
pages 294–5: ibid, pp 91–5
page 296: Naomi Klein, 'Capitalism vs. the Climate', *The Nation*
page 296: Larry Bell, *Climate of Corruption: Politics and Power Behind the Global
Warming Hoax*, Greenleaf Book Group Press, Texas, 2011, p xi
page 297: Naomi Klein, 'Capitalism vs. the Climate', *The Nation*

Chapter 21   Coming Home

page 301: Wendy Bacon et al, 'A Sceptical Climate: Media Coverage of Climate
Change in Australia 2011', Parts 1 and 2, Australian Centre for Independent
Journalism & Global Environmental Journalism Initiative, University of
Technology Sydney, Sydney, 2011, http://datasearch2.uts.edu.au/acij/
investigations/detail.cfm?ItemId=29219

Chapter 22   Going, Going, Gone . . .

page 306: 'UQ Scientist Named Coordinating Lead Author for Next IPCC Report',
University of Queensland News, 24 June 2010, http://uq.edu.au/news/index.
html?article=21390
page 307: 'CO2 Now', National Oceanic and Atmospheric Administration, http://
co2now.org
page 309: Jo Chandler, *Feeling the Heat*, p 83
page 310: Duke University, 'Climate Change Driving Tropical Birds to Higher
Elevations', *Science Daily*, 8 December 2011, www.sciencedaily.com/releases/
2011/12/111208121028.htm
page 310: Richard Alleyne, 'Climate Change Could Destroy 80 per cent of
Rainforest by Next Century', *Telegraph* (UK), 5 August 2010, www.telegraph.
co.uk/earth/earthnews/7928296/Climate-change-could-destroy-80-per-cent-
of-rainforest-by-next-century.html
page 311: Food and Agriculture Organization of the United Nations, 'Climate
Change and Food Security in Pacific Island Countries', Rome, 2008, p 8,
ftp://ftp.fao.org/docrep/fao/011/i0530e/i0530e.pdf
pages 311–12: Cheryl Jones, 'Frank Fenner Sees No Hope for Humans', *The
Australian*, 16 June 2011, www.theaustralian.com.au/higher-education/frank-
fenner-sees-no-hope-for-humans/story-e6frgcjx-1225880091722
pages 312–13: Jessica Marshall, 'Global Warming May Cook Sea Turtle Eggs:
Turtle Breeding Grounds Could Eventually Become Makeshift Ovens',
Discovery News, 2 February 2010, http://news.discovery.com/animals/
sea-turtles-global-warming.html

Chapter 23   Line in the Sand

page 320: Paddy Manning, 'The Climate Deadline is Closer Than We Think',
Sydney Morning Herald, 12 November 2011, www.smh.com.au/business/
the-climate-deadline-is-closer-than-we-think-20111111–1nbhp.html

Epilogue

page 332: Matt McDermott, 'Climate Change, Like Slavery, Needs a True
Cultural Shift to Stop It', Treehugger, 29 October 2010, www.treehugger.com/
corporate-responsibility/climate-change-like-slavery-needs-a-true-cultural-
shift-to-stop-it.html

page 332: Penny Wong, cited in Senate Hansard No 14, 2011, Forty-Third
Parliament: First Session—Fourth Period, Commonwealth of Australia,
p 8500, www.financeminister.gov.au/speeches/2011/sp_081111.html

page 333: Piers Akerman, 'The New Dark Age Begins Today', Telegraph, 8
November 2011, http://blogs.news.com.au/dailytelegraph/piersakerman/
index.php/dailytelegraph/comments/the_new_dark_age_begins_today/

page 334: Barry O'Farrell, quoted in AAP, 'Thousands Cut off as Floodwaters Peak
in Northern NSW Town of Moree', The Australian, 3 February 2012, www.
theaustralian.com.au/news/nation/thousands-isolated-evacuated-in-near-
record-nsw-floods/story-e6frg6nf-1226261475324

ACKNOWLEDGMENTS

WRITING OVER 90,000 WORDS IN TWO AND A HALF MONTHS TURNED out to be slightly harder than I'd initially expected. I am hugely grateful to everyone who made it possible. Simon, thank you for allowing our honeymoon in Byron Bay to be turned into a 'writer's retreat' and patiently watching the cricket day after day as I typed the first chapters. Without all your support, the book simply couldn't have been written—and our kitchen would be a complete disaster zone. I love you always.

I am incredibly lucky to have been blessed with a wonderful family. Thank you Mum for always believing in me, and for reading every chapter with your English-teacher's eagle eyes. Dad, thanks for your help trudging through endnotes. Uncle Geoff—thank you, thank you, thank you for letting us film on the farm. Uncle Bill and Aunty Julie—you have both given me so much support over the years. Sis—your text messages have been great. Nan and Bob—I love you! I wouldn't be who I am today without you. Roslyn and Sarah—you are the future. Keep shifting the power.

Thank you to my agent Gaby Naher for approaching me in 2009 and suggesting I get in touch if I ever decided to write a book. I'm so relieved you remembered me when I emailed you out of the blue two years later to tell you it was time! Straight away, you asked me the most important question: why do you want to write this book? And then you introduced me to a powerhouse editor—Colette Vella from Melbourne

University Press. Colette, thank you for taking on this project in such an unusually short timeline, for understanding its vision, and for being the most 'can do' editor I could imagine. Also at Melbourne University Press, many thanks to Louise Adler, Jacqui Gray, Georgie Bain and Terri King. You guys are awesome. Others who contributed to the book process include Simon Paterson, Samantha Collins, Meryl Potter, Josh Durham and Ian Faulkner.

Madlands would never have been completed without my friends and fellow writers who went above and beyond in their advice. Rose Powell, your edits made the book a gazillion times better. You are an incredibly powerful writer and editor. Rebecca Giggs, many thanks for your thoughtful counsel: it was spot-on. You are both far better writers than me, and I can't wait to read your books! To others who helped but don't want to be named: you know how important you are to me, and to this book.

This journey overlapped with the planning of our wedding, and I can't thank enough the people who helped. Your efforts meant I could focus less on dresses, flowers and logistics and more on the documentary and book. To all the Make Believe crew, thanks for your unfailing support and for allowing me the time to pursue this project. To Grace Mang, Adrienne Ryan and Angeline Meloche: thank you for your encouragement over the past few months. Amanda McKenzie, your help preparing for the documentary was extraordinary and I am so lucky to have you as a friend.

Australia is blessed with a group of extremely talented climate scientists. I owe many of them an enormous debt for patiently answering my endless questions about the science. Huge thanks to Professor Matthew England and Professor Steven Sherwood from UNSW, Professor Ove Hoegh-Guldberg from the University of Queensland, Professor David Karoly from the University of Melbourne, Professor Dave Griggs from Monash University and Dr Mark Howden from the CSIRO for your generous assistance. I am forever grateful to John Cook from the University of Queensland and his colleagues at Skeptical

Science. John, your advice has been invaluable. Thank you for being so generous with your time, and for the illustrations you re-created from the peer-reviewed science.

Gigantic thanks to all the experts who gave their time to meet with us as part of the documentary journey. In addition to the scientists, special thanks go to Professor Naomi Oreskes, Rear Admiral David Titley, Zac Goldsmith and Irene Khan for the compassionate, intelligent contributions to discussions about climate change. Thank you also to Jo Chandler and Bill McKibben for writing your books. *Feeling the Heat* and *Eaarth* made me realise it was possible to write a different kind of climate change book.

I am in awe of Ellen Sandell and the rest of the Australian Youth Climate Coalition team. You are all mighty campaigners and make me immensely proud to be chair of such an innovative, passionate and mission-driven organisation. Like Paul Hawken said: 'When asked if I am pessimistic or optimistic about the future, my answer is always the same: If you look at the science about what is happening on earth and aren't pessimistic, you don't understand data. But if you meet the people who are working to restore this earth and the lives of the poor, and you aren't optimistic, you haven't got a pulse.' Keep demanding the impossible, because we have to avoid the unimaginable. And remember: those who say it can't be done should get out of the way of those already doing it!

Finally, thank you to Nick Minchin for being open-minded enough to travel around the world with me, and for being such a good sport about being included in this book. I truly hope that one day you'll join me in focusing your formidable talents towards shifting Australia's economy towards a clean energy future.